Electric Circuits

ELECTRIC CIRCUITS

Nihal Kularatna

Oxford Book Company
Jaipur ▪ New Delhi

Published By

Oxford Book Company

Regd. Off. "Bony Residency", Gate No. 2
Opp. Tilak Public School, Vishweswariya Nagar,
Gopalpura Road, Jaipur - 302018, India
Ph: 0141 - 2761280, Telefax: 2761381
email - oxfordbookjpr@gmail.com
website - oxfordbookcompany.com

Delhi Office: 102, 1st Floor, Satyam House,
4327/3, Ansari Road, Darya Ganj,
New Delhi - 110002, India
Ph: 011-45652440

Published in India, 2025

ISBN : 9789380179377

Price : ₹ 2195/-

Typeset by Shivangi Computers, Jaipur

Printed and bound in India by Thomson Press Ltd.,
New Delhi

Preface

Electronics is probably one of the few subjects where the "knowledge halflife" of a professional is very short. Today it is probably only 3 to 4 years.

Designing electronic systems today requires a unique combination of (1) fundamentals; (2) research and development directions in the latest semiconductors and passive components; (3) nitty-gritty aspects within the "mixed signal world;" (4) access to component manufacturers' data sheets, design guidelines, and development environments; and, most importantly, (5) a timely and practical approach to overall aspects of a design project.

This work attempts to address several areas of analog and mixed signal design, including power supply design, signal conditioning, essentials of data conversion, and signal processing, while summarizing a large amount of information from theory texts, application notes, design bulletins, research papers, and technology magazine articles. In a few chapters I had the assistance of experts in different subject areas, as chapter authors.

Nihal Kularatna

Contents

Chapter 1

Introduction

For an adequate understanding of the subject matter of this book, it is essential not only that the text be read and understood but especially that the reader complete all of the problems. The importance of solving the problems cannot be overstated. Some of the points in the text will seem quite obvious, and it is hoped that all of them will seem logical and understandable. However, it has been the author's repeated experience in teaching this material that people may feel that they understand it completely and yet are unable to put into practice even some of the simpler principles. If the reader really has no need to solve the problems in order to understand them, then the problems will be very easy for him. If the reader finds them hard, then they must have been needed.

The general method of presentation in this book is to put basic principles and practical applications in close juxtaposition. It is certainly true that an understanding of such principles is essential to the design of new circuits, but, at the same time, understanding is not always sufficient for the successful application of these principles. Frequently, the applications must be specifically explained.

A complete mastery of the concept of electric resistance helps very little in building a piece of apparatus if the builder does not know what a resistor looks like or what to do about the wires sticking out of each end. The following pages, therefore, contain almost as many pictures of actual circuit parts as circuit diagrams, and as much concrete description of elements which go into electric apparatus as discussion of the properties of electricity flowing through conductors.

The examples discussed and the problems to be solved are drawn primarily from the behavioural sciences. Some of the most common circuit problems will be discussed specifically, so that the reader may use the solutions presented directly if he insists. However, this book is not intended as a handbook of solutions to common circuit problems and will not serve that function very well. Many of the problems given require the building of actual electric circuits, and a note about shock hazards might be in order here. None of the circuits to be built at the beginning of this book requires more than 12 volts to operate it. It is virtually impossible to get a real shock from these circuits. In fact, except for very few circuits which will be specifically pointed out, it is difficult even to feel a shock when trying very hard to do so. There is no need, therefore, to fear electrocution; each circuit capable of giving any shock at all will be so identified. It will be necessary to borrow or buy a number of circuit parts in order to solve the problems presented. These parts can be bought at any radio supply store, but they are all common enough so that many of them can probably be found around the laboratory.

A list of these parts follows:

- A 100-foot spool of solid (not stranded) hookup wire (#22 solid-type MW plastic-insulated wire, or equivalent).
- Ten feet of lamp cord.
- Ten alligator clips (screw connection).
- A 1-pound spool of rosin core solder.
- Four terminal strips (at least four terminals each), lugtype, for soldering. To be screwed to board.
- Four pilot light bulbs (6 volts, type 47).
- Two sockets for above light bulbs.
- One battery, Burgess-type 5156 "B or C" (3. 4½, 6, 9, 10½, 16½, 22½ volts) clip terminals (or equivalent).
- One male wall plug.
- Two resistors (5 watts, 30 ohms each).
- One potentiometer (10 watts, 50 ohms, wire-wound).
- One potentiometer (4 watts, 5000 ohms, wire-wound).
- Two set screw knobs for potentiometers.
- Two capacitors (at least 150 microfarads, 50 volts, electrolytic).
- Two microswitches, SPDT.
- One toggle switch, DPDT.
- One relay, 115 volts a-c, at least 4PDT, not enclosed (with visible working parts).
- One relay, SPDT, sensitive (e.g. with at least 1000-ohm coil resistance, closes on 10 volts or less, closes on 10 milli amperes or less), not enclosed (with visible working parts).
- One toy motor (3 volts).

If none of these items can be borrowed, they should not cost more than a total of about $25.00, and may be much cheaper at surplus stores. The following tools should be available:

- A medium-sized screw driver.
- A small soldering iron.
- One pair of long-nosed pliers with wire cutters or, preferably, one pair of long-nosed pliers and a separate pair of wire cutters.

- One ohm-volt-ammeter.

It would be most efficient if all these items were on hand before the book proper is begun rather than trying to locate each part or tool when it is first required. Many of the items listed above should be mounted on a board for convenience in setting up circuits.

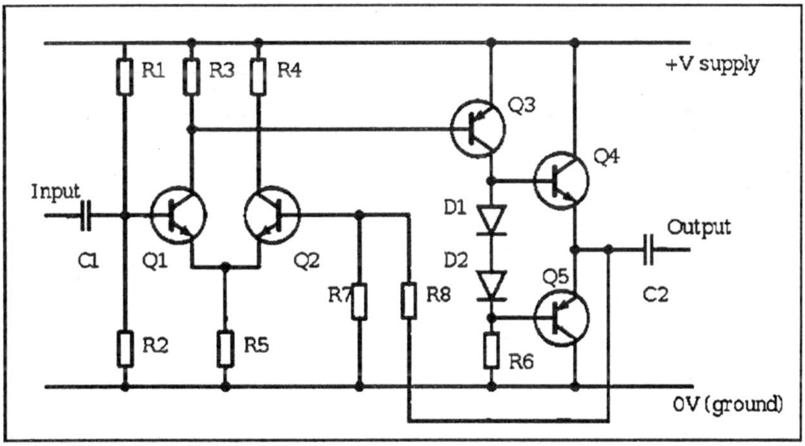

Fig. A Board with Mounted Circuit Elements.

Find a piece of wood about one foot square, and mount on it each of the components listed below.

- All four terminal strips.
- Two of the flashlight bulbs in their sockets.
- Battery.
- Both potentiometers (variable resistors).
- Both microswitches.
- Toggle switch.
- Both relays.
- Motor.

The particular locations of each of these components are not important. To speed up construction of some of the simpler circuits, "clip leads" are convenient. A clip lead is a wire with an alligator clip at each end. Make five clip leads by connecting alligator clips to each end of five 12 inch-long pieces of hookup wire.

THE CONCEPT OF ELECTRICITY

When you pet a cat on a dry day, both you and the cat get shocked. The reason for this is that there are loosely bonded electrons on the surface of your skin, and with your first stroke some of them are rubbed off onto the cat's fur. The cat now has more electrons than normal, and you have fewer (the cat is negatively charged, you are positive). The excess electrons on the cat's fur redistribute themselves so that the charges are most dense at the tip of each hair. When you begin the next stroke, some of these electrons jump back to your hand, illustrating the fact that electrons are repelled by other electrons, and attracted by atoms that have fewer electrons than normal. The sparks that you feel and hear (and can see in a dark room) are groups of electrons jumping the gap between the tips of the cat's hairs and your hand.

CURRENT AND VOLTAGE

If you were to connect a wire between a petted cat and the hand that did the petting, electrons would flow through the wire from the cat to the hand. This flow of electrons is called electric *current*. The more electrons flowing past any point in the wire in a given time, the higher the current. If more electrons are rubbed off in the first place, the "charge" on the cat and the hand is higher, and the tendency for electrons to flow in the wire is greater. The "tendency" for current to flow between two points is called the *voltage* between the two points. From these crude definitions, it is clear that the higher the voltage, the greater the current, other conditions being equal.

Suppose that there were a means of continuously supplying a voltage between two points, and that the two points were connected together by a wire; then current would continuously flow in the wire. But it will lead to some confusion later if you picture current as the flow of a series of individual electrons from one end of the wire to the other. What actually happens is this. Any good conductor of electricity, e.g., the wire, contains many loosely bonded electrons. When an electron enters one end of the wire, it displaces a second electron from one of the nearby atoms. This second electron displaces a third, and so on. This means that, when one electron enters one end of a wire, another, different electron leaves the other end.

Electric current is conducted along wire in a manner very analogous to the way in which a sound wave is conducted through the air. The energy in an electric current travels much faster than do the individual electrons. Individual electrons in a conductor travel in the order of magnitude of 100 feet per second, but the energy itself travels so fast that conduction in essentially all circuits can be considered instantaneous. This is one reason why it is usually better to think of electric current in a conductor as a flow of electric energy rather than of electrons themselves.

Cat fur is not a particularly convenient supplier of electric energy. The sources you will use most frequently are batteries and wall sockets. Since batteries are easier to understand and involve many of the same principles as wall sockets, batteries will be discussed first.

BATTERIES

There are many different types of batteries. The kind you are probably most familiar with is the one used in flashlights. It is one of a type called dry cells, because it contains no liquids. It consists of a zinc can containing chemicals and a carbon rod. There is usually a paper jacket over the can, except at the bottom where the zinc is exposed, and at the top where the end of the carbon is covered by a brass shell. The chemicals operate on the can and the carbon rod in such a way as to generate voltage between them.

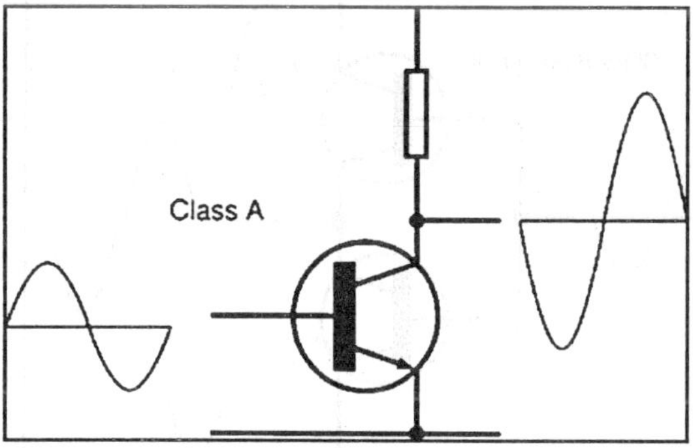

Fig. Cross Section Through a Typical Flashlight

That is, the zinc has an excess of electrons (it is negative) and the carbon has fewer electrons than normal (it is positive). If a wire were connected between the carbon rod and the can, current would flow through the wire until the battery ran down. A flashlight bulb contains just such a wire. Current running through the wire heats it up to the point where it emits light (the wire, or filament, is enveloped in an inert gas so that it does not burn up). The terminals of flashlight batteries are such that it is difficult to make connections directly to them. When such batteries are used, they are typically mounted in holders. But dry-cell batteries working on identical principles do come in more versatile packages. The zinc can and the carbon rod are connected to screw terminals I (1), clips (2), or a socket (3), so that they are easier to connect to other circuit elements. Dry-cell batteries can be obtained to give voltages anywhere from 1.5 volts to 300 volts, or more. For the moment, we will consider only the 1.5-volt type.

RESISTANCE

Suppose you were to connect a wire from one terminal of a 1.5volt battery to the other. The wire has a property known as electric *resistance*. Resistance means just what you would intuit it to mean. The more resistance a conductor has, the harder it is for current to get through it—fewer loosely bonded electrons are available. When a wire is connected between the terminals of the battery, and if the wire has a higher resistance, less current will flow.

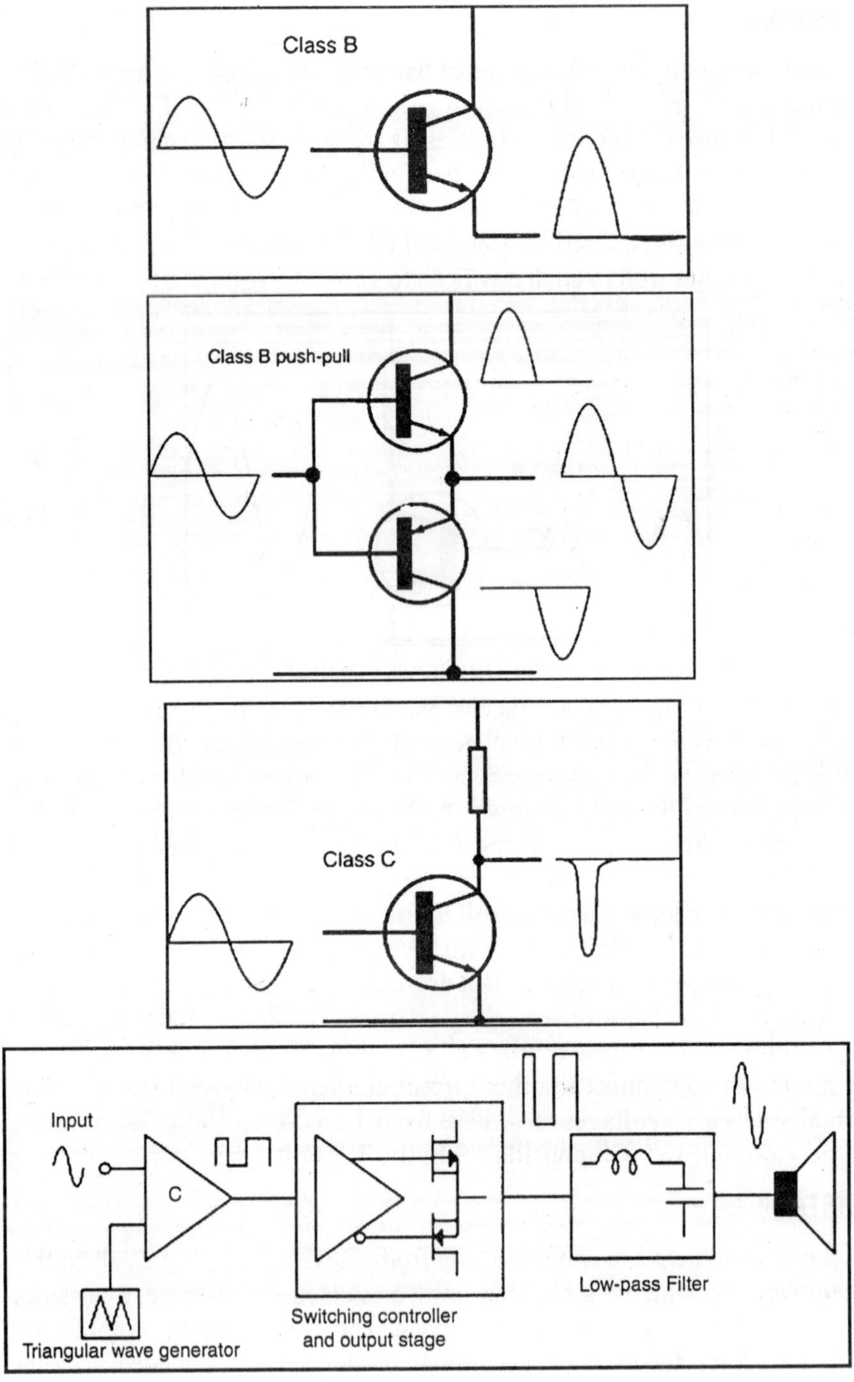

Fig. (*a*) Holders for Flashlight Batteries. (*b*) Various batteries; (1) with Screw Terminals; (2) Clips; and (3) a Socket.

An ordinary wire has a very low resistance, so that if you actually did connect it from one terminal of the battery to the other, so much current would flow that the chemicals in the battery would be exhausted very rapidly.

The electrical resistance of an object is a measure of its opposition to the passage of a steady electrical current. An object of uniform cross section will have a resistance proportional to its length and inversely proportional to its cross-sectional area, and proportional to the resistivity of the material.

Discovered by Georg Ohm in the late 1820s, electrical resistance shares some conceptual parallels with the mechanical notion of friction. The SI unit of electrical resistance is the ohm, symbol Ω. Resistance's reciprocal quantity is electrical conductance measured in siemens, symbol S.

The resistance of a resistive object determines the amount of current through the object for a given potential difference across the object, in accordance with Ohm's law:

$$I = \frac{V}{R}$$

where

R is the resistance of the object, measured in ohms, equivalent to $J \cdot s/C^2$

V is the potential difference across the object, measured in volts

I is the current through the object, measured in amperes

For a wide variety of materials and conditions, the electrical resistance does not depend on the amount of current through or the amount of voltage across the object, meaning that the resistance R is constant for the given temperature. Therefore, the resistance of an object can be defined as the ratio of voltage to current:

$$R = \frac{V}{I}$$

In the case of nonlinear objects (not purely resistive, or not obeying Ohm's law), this ratio can change as current or voltage changes; the ratio taken at any particular point, the inverse slope of a chord to an I–V curve, is sometimes referred to as a "chordal resistance" or "static resistance".

DC Resistance

The resistance R of a conductor of uniform cross section can be computed as

$$R = \frac{\ell \cdot \rho}{A}$$

where

ℓ is the length of the conductor, measured in meters

A is the cross-sectional area, measured in square meters

ρ (Greek: rho) is the electrical resistivity (also called *specific electrical resistance*) of the material, measured in Ohm · meter. Resistivity is a measure of the material's ability to oppose electric current.

For practical reasons, any connections to a real conductor will almost certainly mean the current density is not totally uniform. However, this formula still provides a good approximation for long thin conductors such as wires.

AC Resistance

If a wire conducts high-frequency alternating current then the effective cross sectional area of the wire is reduced because of the skin effect. If several conductors are together, then due to proximity effect, the effective resistance of each is higher than if that conductor were alone.

Causes of Resistance

In Metals

A metal consists of a lattice of atoms, each with a shell of electrons. This can also be known as a positive ionic lattice. The outer electrons are free to dissociate from their parent atoms and travel through the lattice, creating a 'sea' of electrons, making the metal a conductor. When an electrical potential difference (a voltage) is applied across the metal, the electrons drift from one end of the conductor to the other under the influence of the electric field.

Near room temperatures, the thermal motion of ions is the primary source of scattering of electrons (due to destructive interference of free electron waves on non-correlating potentials of ions), and is thus the prime cause of metal resistance. Imperfections of lattice also contribute into resistance, although their contribution in pure metals is negligible.

The larger the cross-sectional area of the conductor, the more electrons is available to carry the current, so the lower the resistance. The longer the conductor, the more scattering events occur in each electron's path through the material, so the higher the resistance. Different materials also affect the resistance.

In Semiconductors and Insulators

In metals, the Fermi level lies in the conduction band giving rise to free conduction electrons. However, in semiconductors the position of the Fermi level is within the band gap, approximately half-way between the conduction band minimum and valence band maximum for intrinsic (undoped) semiconductors. This means that at 0 Kelvin, there are no free conduction electrons and the resistance is infinite. However, the resistance will continue to decrease as the charge carrier density in the conduction band increases.

In extrinsic (doped) semiconductors, dopant atoms increase the majority charge carrier concentration by donating electrons to the conduction band or accepting holes in the valence band. For both types of donor or acceptor atoms, increasing the dopant density leads to a reduction in the resistance. Highly doped semiconductors hence behave metallic. At very high temperatures, the contribution of thermally generated carriers will dominate over the contribution from dopant atoms and the resistance will decrease exponentially with temperature.

In ionic Liquids/Electrolytes

In electrolytes, electrical conduction happens not by band electrons or holes, but by full atomic species (ions) traveling, each carrying an electrical charge. The resistivity of ionic liquids varies tremendously by the concentration - while distilled water is almost an insulator, salt water is a very efficient electrical conductor. In biological membranes, currents are carried by ionic salts. Small holes in the membranes, called ion channels, are selective to specific ions and determine the membrane resistance.

Resistivity of Various Materials

Band theory Simplified

Quantum mechanics states that the energy of an electron in an atom cannot be any arbitrary value. Rather, there are fixed energy levels which the electrons can occupy, and values in between these levels are impossible. The energy levels are grouped into two bands: the valence band and the conduction band (the latter is generally above the former). Electrons in the conduction band may move freely throughout the substance in the presence of an electrical field.

In insulators and semiconductors, the atoms in the substance influence each other so that between the valence band and the conduction band there exists a forbidden band of energy levels, which the electrons cannot occupy. In order for a current to flow, a relatively large amount of energy must be furnished to an electron for it to leap across this forbidden gap and into the conduction band. Thus, even large voltages can yield relatively small currents.

Differential Resistance

When resistance may depend on voltage and current, differential resistance, incremental resistance or slope resistance is defined as the slope of the *V-I* graph at a particular point, thus:

$$R = \frac{dV}{dI}$$

This quantity is sometimes called simply *resistance*, although the two definitions are equivalent only for an ohmic component such as an ideal resistor. For example, a diode is a circuit element for which the resistance depends on the applied voltage or current.

If the *V-I* graph is not monotonic (i.e. it has a peak or a trough), the differential resistance will be negative for some values of voltage and current. This property is often known as *negative resistance*, although it is more correctly called *negative differential resistance*, since the absolute resistance *V/I* is still positive. Example of such an element is a tunnel diode.

Temperature-Dependence

Near room temperature, the electric resistance of a typical metal increases linearly with rising temperature, while the electrical resistance of a typical semiconductor decreases with rising temperature. The amount of that change in resistance can be calculated using the temperature coefficient of resistivity of the material.

At lower temperatures (less than the Debye temperature), the resistance of a metal decreases as T^5 due to the electrons scattering off of phonons. At even lower temperatures, the dominant scattering mechanism for electrons is other electrons, and the resistance decreases as T^2.

At some point, the impurities in the metal will dominate the behaviour of the electrical resistance which causes it to saturate to a constant value. Matthiessen's Rule (first formulated by Augustus Matthiessen in the 1860s; the equation below gives its modern form) says that all of these different behaviors can be summed up to get the total resistance as a function of temperature,

$$R = R_{\text{imp}} + a\text{T}^2 + bT^5 + cT$$

where R_{imp} is the temperature independent electrical resistivity due to impurities, and *a*, *b*, and *c* are coefficients which depend upon the metal's properties. This rule can be seen as the motivation to Heike Kamerlingh Onnes's experiments that lead in 1911 to discovery of superconductivity.

The electric resistance of a typical intrinsic (non doped) semiconductor decreases exponentially with the temperature:

$$R = R_0 e^{-aT}$$

Extrinsic (doped) semiconductors have a far more complicated temperature profile. As temperature increases starting from absolute zero they first decrease steeply in resistance as the carriers leave the donors or acceptors. After most of the donors or acceptors have lost their carriers the resistance starts to increase again slightly due to the reducing mobility of carriers (much as in a metal). At higher temperatures it will behave like intrinsic semiconductors as the carriers from the donors/acceptors become insignificant compared to the thermally generated carriers.

The electric resistance of electrolytes and insulators is highly nonlinear, and case by case dependent, therefore no generalized equations are given.

Measuring Resistance

An instrument for measuring resistance is called an ohmmeter. Simple ohmmeters cannot measure low resistances accurately because the resistance of their measuring leads causes a voltage drop that interferes with the measurement, so more accurate devices use four-terminal sensing.

A SIMPLE CIRCUIT

In order to get a feel for what voltage, current, and resistance mean, it is worthwhile to set up a simple electric circuit. You will need three clip leads, one light bulb and socket, the 50 ohm variable resistor (potentiometer), and the battery. The circuit can be schematically represented by a diagram.

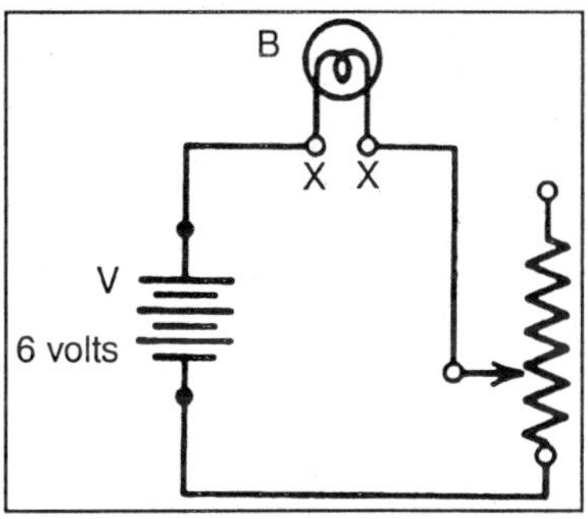

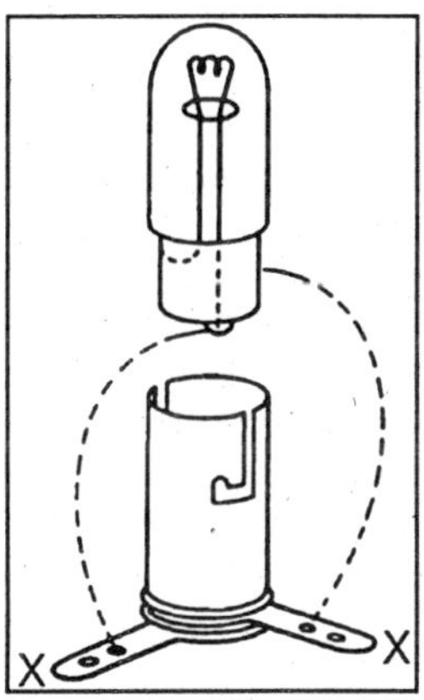

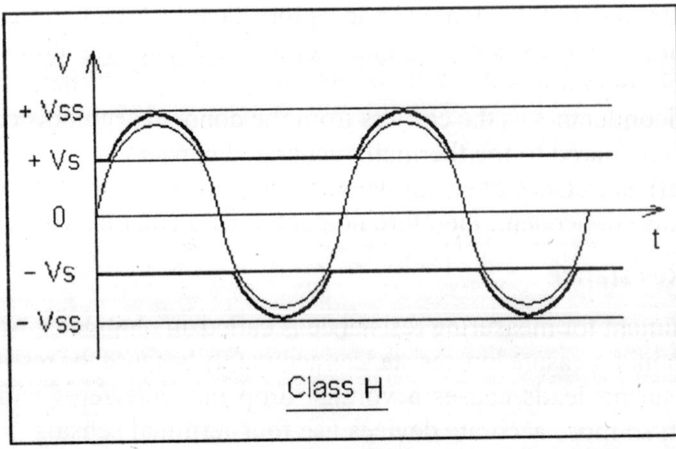

Fig. (*a*) Circuit Diagram for Varying the Brightness of a Light Bulb. *V* Represents the Battery, *B* the Bulb, and the zig-zag Line a Variable Fifty Ohm Resistor. (*b*) The Internal Connections of the Bulb and Socket. (*c*) The Anatomy of the Variable Resistor (Potentiometer).

In this diagram straight lines represent connections (by clip leads), and the series of short lines labeled *V* represents the battery.

The figure enclosed in the circle, labeled *B*, represents the light bulb and its socket. Voltage impressed between the two metal tabs, or terminals, on the socket causes current to flow in one terminal, through the lead-in wires and the lamp filament, and out the other terminal.

The zig-zag line represents the resistor. Examine the resistor. If it has a cover on it, take the cover off so that you can see the parts inside. The resistive element itself is a long wire wrapped around a cylindrical core. There are three metal tabs on the outside of the resistor. The two outside tabs are connected to the two ends of this resistance wire. The middle tab is connected to an arm called the slider, or wiper, which slides along the resistance coil. On the circuit diagram, the two outside tabs are represented by the two ends of the zig-zag, and the middle tab—the slider—is represented by the arrow.

Connect up the circuit. If you have done this correctly, the brightness and colour of the light bulb will change as the slider of the resistor is moved. Now consider why this should be. The actual resistance through which current must flow is the resistance of the light bulb plus that of the wire between one end of the resistor and the slider. When the slider is moved from one end to the other, the effective length of resistance wire changes, and so the amount of resistance in the circuit changes. Because the voltage generated in the battery is constant, that is, the "tendency" for current to flow in the circuit does not change, the amount of current actually flowing will decrease when the amount of resistance increases. The brightness of the bulb decreases correspondingly when the current through it decreases.

This circuit is often used to control the brightness of slide projectors, house lights, etc. (although it should be noted that the colour is changed along with brightness, which is undesirable under many experimental conditions). The only difference between your circuit and one that might be used to control the intensity of a projected subliminal stimulus, for example, is that a physically bigger resistor would be needed to handle the large current of a projector bulb.

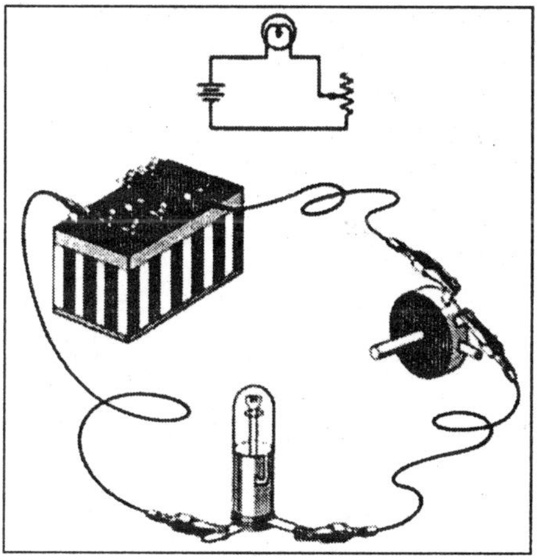

Fig. The Connections as Represented in the Circuit Diagram of Figure. That Diagram is Redrawn at the Top of This Figure.

Be sure that you can predict, by looking at the resistor, whether the bulb should get brighter or dimmer as the slider is turned, say, clockwise. Then change the circuit so that the opposite will be true. Do not dismantle the circuit yet (but disconnect one of the wires to the battery, so that it will not run down).

METERS

In the simple circuit you have just built, the light bulb acts as a sort of meter, crudely indicating the amount of current flowing in the circuit. A device which measures, quantitatively, the amount of current flowing in a circuit is called an ampere meter, or *ammeter*. It is called this because, as will be discussed later, electric current is commonly measured in units called amperes. It is important for anyone working with electric apparatus to understand the fundamental principles underlying the operation of ammeters.

Principles of Meter Operation

When electrons flow through a wire, they cause the wire to heat up. Light bulbs and electric heaters are examples of this. Current flowing through a wire also generates a magnetic field around the wire. If current flows through a wire that is wound into a coil, a magnetic field will be formed around the entire coil. As the current through the coil is increased, the strength of the magnetic field is increased in direct proportion.

If two magnetic fields are juxtaposed, they will either attract or repel each other, depending upon the relationship of their polarities, and this effect forms the basis for the actions of almost all ammeters. The typical ammeter consists of a permanent magnet and a small coil of wire which pivots. The coil is mounted in the magnetic field of the permanent magnet and is held in position by springs.

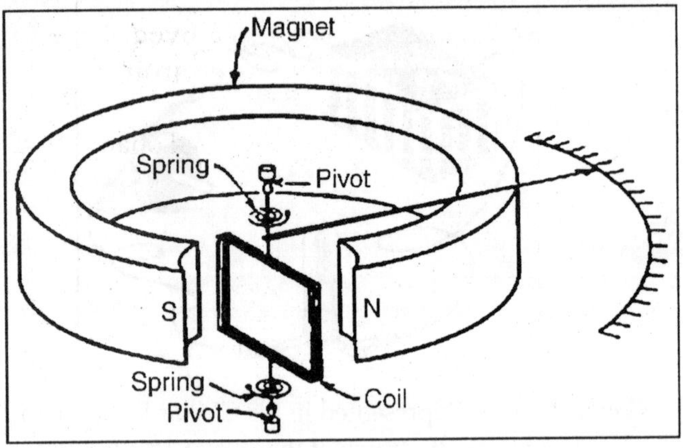

Fig. Essential Parts of a Typical Meter Movement

When current is passed through the coil, a magnetic field is generated around the coil. This field is pushed at one end and pulled at the other by the field of the permanent magnet, and the coil therefore turns on its pivots. As the coil turns farther and farther, the springs exert an increasing force in the opposite direction; that is, they tend more strongly to return the coil to its original position. The coil moves until the magnetic and spring forces are equal. The higher the current through the coil, the stronger will be the magnetic forces, and the farther the coil will turn.

A pointer which moves across a scale is connected rigidly to the coil. Thus the strength of the current can be read by the position of the pointer on the scale. The two ends of the coil are connected to terminals on the outside of the meter so that any current to be measured may be passed through the coil. To determine how much current is flowing in some circuit, the ammeter must be connected into the circuit in such a way that either all or some known proportion of the unknown current flows through the meter coil. Meters used to measure voltage (voltmeters) and resistance (ohmmeters) operate on exactly the same principles.

In fact, a single meter movement, that is, magnet and coil, can be used to measure current, voltage, or resistance. You need only to change the circuit external to the movement itself in order to accomplish this. The meter that you have contains one movement and the switches on the front panel connect the appropriate built-in circuits to it. These conversion circuits are important and will be explained later.

Use of the "Multimeter"

Set your meter so that it will read d-c current, 500 milliamperes (ma) or more, full scale. Your meter should have two wires plugged into it. These are called test leads and are used to connect the meter into the circuit. Connect the meter in your light-bulb circuit. The reading on the meter will change as the slider is moved. Note from the circuit configuration that all the current flowing through the bulb must also pass through the meter.

Now remove the meter from the circuit and change the settings to read on a scale of at least 6 volts, d-c. Reconnect the circuit, using the voltmeter. The meter now reads the voltage being put out by the battery. Notice that the voltage does not change when the resistance slider is moved.

Remove the meter again and change the settings so that it reads ohms × 1. The meter is now internally connected to read resistance in the standard units, ohms. Inside the meter case there are several batteries and they have now been connected to the coil and test leads so that some of the voltage from the batteries appears between the test leads. Touch the two leads together. If the meter is working properly, the needle will swing up toward the full-scale reading which, on the ohms scale, is *zero* ohms. There is a knob on the meter that says something like "ohms adjust." Turn this knob until the meter reads zero ohms when the test leads are touching each other.

This procedure assumes that the resistance of the test leads themselves is zero, and for all practical purposes it is. If the needle will not go all the way to zero (full scale), the meter needs new batteries. Now *make sure that your battery is disconnected* from your 50 ohm variable resistor, and connect the test leads between the slider and one end of the resistor. Read the resistance from the meter for several settings of the slider. The ohm scale on the meter is the reverse of the volt and ampere scales. When the resistance is zero, the battery in the meter drives enough current through the meter coil to cause it to swing to full scale. The "ohms adjust" knob gives this current the proper value.

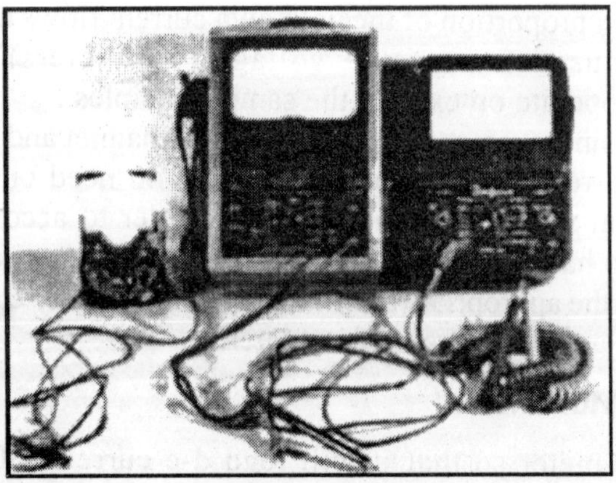

Fig. Three Typical Volt-ohm-ammeters Set to Read at Least 500 Milliamperes d-c Full Scale. This Setting may be made with a Single Knob on some Meters, by two Knobs on Others (One for 500 Milliamperes and Another for d-c), or, on Still Others, by a Combination of a Knob Setting and the placement of the two test leads in the appropriate sockets.

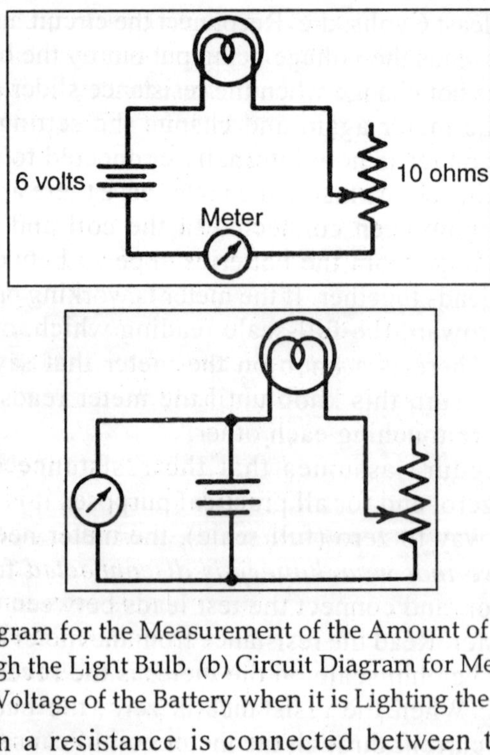

Fig. Circuit Diagram for the Measurement of the Amount of Current Flowing Through the Light Bulb. (b) Circuit Diagram for Measuring the Voltage of the Battery when it is Lighting the Bulb.

Now, when a resistance is connected between the test leads, it reduces the amount of current that can flow through the meter so that the needle swings back toward less current, indicating more resistance. (The light bulb in your first circuit indicated the amount of current and, at the same time, the amount of resistance, much as an ohmmeter would. The dimmer the bulb, the higher the resistance.) The principles of operation of meters, are well worth mastering. It is certainly possible to use meters without understanding how they operate, but it is safer, both for the meters and for the interpretation of their readings, if the basic principles are understood.

Examples of common misuses of meters will be given in several other sections of this book. At present, however, it is worth mentioning a few aspects of meter use that may not be obvious. It can do a meter a great deal of harm to have it set on the wrong scale when it is connected into the circuit. For instance, if it is set on a scale whose maximum reading is I volt, and you connect it across 6 volts, two things can happen to damage the meter. First, the needle will swing up to full scale very quickly and bang against the stop at that end of the scale. This can put a permanent bend in the needle. A second, more serious, consequence is that more current will flow through the meter coil than it can safely carry and the coil may burn out.

In general, always make sure before you connect a meter into a circuit that it is set to a scale higher than the maximum possible value of whatever is to be measured. It does a meter no harm to be run backwards so long as it is set on a safe scale.

For example, set the meter to read at least 6 volts full scale and connect the test leads to the battery, one to the terminal marked "+," and the other to the one marked "4½v." If you have done this correctly, the meter will read 4.5 volts. If you have the test lead connections reversed, the meter needle will read *less than zero*. This reversed condition will not hurt the meter. Therefore, if you do not know in advance which lead to connect to which terminal, just try out either combination or the right one will give you a reasonable reading.

Never try to measure the current in a battery. You may, out of idle curiosity, wonder how much current there is in your 22½-volt battery, so you may set your meter to read amperes and connect the test leads, one to each side of the battery. This will cause the meter to burn out immediately. What you have *actually* asked the meter is, "How much current will flow when the meter is connected across the battery?" and, since an ammeter has very little resistance to current flow, an enormous amount of current will begin to flow, more than the meter is built to handle. The question of how much current is in a battery is an unanswerable one, not just technically, but logically. It is like asking how fast the water in the Pacific Ocean will flow.

Table: Electrical Units and Symbols

Name	Circuit Symbol	Common Units	Unit Symbol or Abbreviation
Voltage	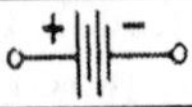	volt	v
		millivolt (10-3volt)	mv
		microvolt (10-6)	?v
		kilovolts (103 volts)	kv
Current	none	ampere	amp or a
		milliampere(10-3 ampere)	ma
		microampere(10-6)	?a
Resistance	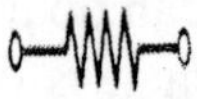		
	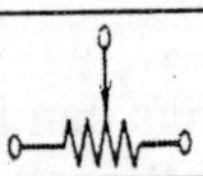		
	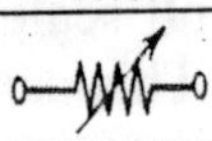		

Ohm's Law and Power

There are two formulas presented in this chapter, and it is essential that both be mastered. The first is Ohm's law and it will be discussed now; the second, the calculation of power, will be discussed later in the chapter.

OHM'S LAW

What is commonly called Ohm's law is the mathematical expression for the relationship already mentioned between the amount of current, voltage, and resistance in any given circuit or part of a circuit. When resistance is constant, the higher the voltage the higher the current. When voltage is field constant, the lower the resistance the higher the current.

That is $I = \dfrac{E}{R}$

where I = the current, in amperes,

E = the voltage, in volts,

R = the resistance, in ohms.

It can be stated in two other equivalent forms:

$E = IR \quad R = \dfrac{E}{I}$

You really need to remember only one of these forms because, knowing one, either of the others is immediately derivable. Be sure to notice that even if all forms are forgotten they can be easily figured out just by realizing the fact that higher voltages and lower resistances give higher currents. The best way to learn the implications of Ohm's law is to solve problems.

Problem 1. Given the circuit, find the current that will flow around it.

Solution. Ohm's law holds when I is in amperes, E in volts, and R in ohms. I = E/R, I = 10 volts/20 ohms = 0.5 ampere.

Problem 2: What must R be changed to in the above circuit in order that 3 amperes flow in the circuit?

Solution. R = E/I = 10/3 = 1/3 ohms.

Problem 3: How large is the voltage across R in the circuit in Problem 1?

Solution. E = IR. I has been found to be 0.5 ampere. Therefore, E = 0.5 × 20 = 10 volts. But looking at the circuit again, it is true by definition that the voltage across R is 10 volts without bothering with Ohm's law. Straight lines in a circuit diagram always represent connections between points. Therefore, the circuit diagram essentially says that the battery is connected across the resistor R. Since there are 10 volts between the terminals of the battery, there must be 10 volts across the resistor.

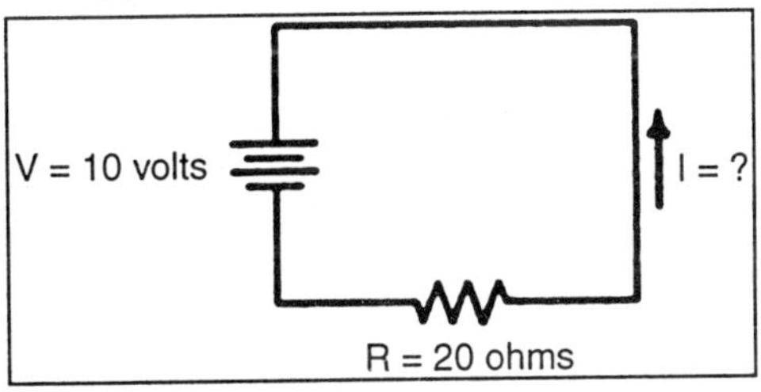

Fig. Circuit Diagram for Problems

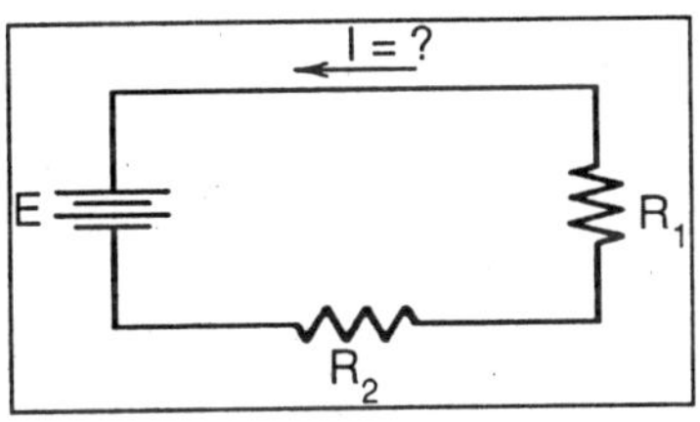

Fig. Circuit Diagram for Problems 1, 2, and 3. 4, 5, and 6.

Problem 4: Solve for the current I in the circuit, given E = 10 volts, R1 = 8 ohms, and R2 = 6 ohms.

Solution. The solution of this problem requires some additional discussion about circuits in general. In the first place, the number of amperes flowing through R1 is the same as through R2, through the connections (represented by the straight lines) and through the battery. If you think about what electric current is, you will see why this must be true. Any given electron entering one end of, say R1, will cause another electron to leave the other end of R1, which, in turn, will cause one electron to enter and another to leave R2, the battery, etc. If more current were to flow in any one part of the circuit than in some other part, electrons would begin to pile up somewhere. This piling up would produce an increase in the voltage at that point, which, in turn, would mean an increase in the tendency for current to flow at that point. And since current and voltage are directly proportional to each other, the resulting increase in current would just balance out the rate of piling up. Thus, the current must be the same through R1, R2, and the battery.

By similar reasoning, the effect of putting two resistors into the circuit, end to end as in this problem, is the same as putting in one resistor whose resistance is the sum of the two separate ones. When resistors are connected end to end, they are said to be in series with each other. The solution of the problem is: I = E/R = 10/(6 + 8) = 10/14 = 0.715 ampere.

Problem 5: E = 10 volts

R1 = 8 ohms

Find the value of R2 which will allow 0.5 ampere to flow in the circuit. $R = \frac{E}{I}$

$8 + x = \frac{10}{0.5}$ $x = 20 - 8 = 12$ ohms

Problem 6: E = 10 volts

R1 = 8 ohms

R2 = 6 ohms

Find the voltage across R1 and the voltage across R2.

Solution. E = IR $I = \frac{10\,\text{volts}}{(6+8)\text{ohms}} = 0.715$ ampere

Therefore

E = 0.715 × 8 = 5.72 volts across R1

E = 0.715 × 6 = 4.28 volts across R2

Note that the voltage across R1 plus that across R2 equals 10 volts. From the circuit diagram this makes sense. This problem illustrates a point which is very important and at the same time difficult for many people really to understand. Whenever a current flows through a resistance, a voltage will appear across that resistance. The magnitude of the voltage will equal the current multiplied by the resistance. Such a voltage has exactly the same mathematical properties as a voltage generated by a battery. Recognition of voltages generated in this way often permits solving problems which otherwise have only very circuitous solutions.

Problem 7: Put together the circuit. Now calculate the voltages across R1 and R2, as in Problem 6, and check your answer by measuring the voltages. Measure the voltage across R1 by setting the meter to read volts on the appropriate scale (at least 6 volts full scale) and then connect the test leads, one to each end of R1. The voltage across R2 is measured in the same way. Next, calculate the amount of current that will flow in the circuit and check your answer by measuring the current.

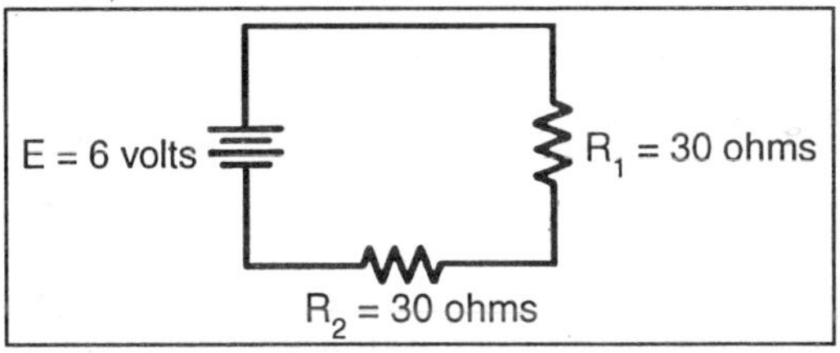

Fig. Circuit to be Constructed for Problem 7.

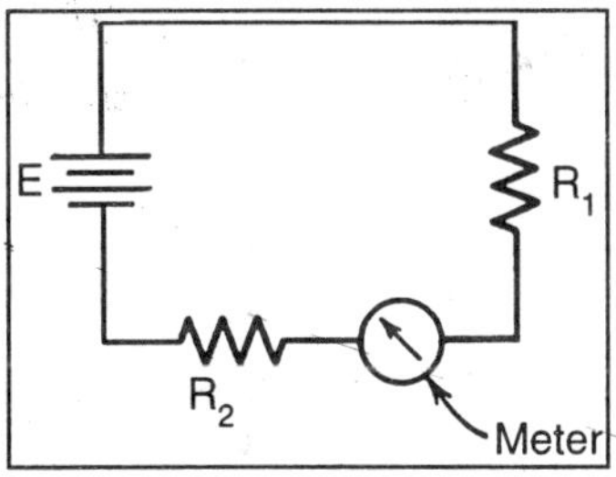

Fig. Circuit Diagram for the Measurement of the Amount of Current Flowing in the Circuit Given in Problem 7.

Set the meter to read d-c amperes on the appropriate scale (at least 500 milliamperes full scale). Then disconnect the circuit any place and connect it back together again through the meter.

POWER

The rate at which a system does work is called the power of the system. In electric systems, power is used to light lamps, run motors, heat resistors, shock rats, etc. The amount of power is usually measured in units called watts (or milliwatts, or microwatts). The formula for the calculation of *watts* should be learned. It is as follows:

Power (in watts) = E (in volts) × I (in amperes)

$$W = E \times I$$

When part of an electric circuit is damaged because of improper circuit connections, the damage can almost always be attributed to heat, e.g., a misconnected meter may literally "burn out." So will a light bulb run at too high a voltage. The amount of power that a circuit element can consume without burning up may be specified in watts. Some circuit elements, for example, some resistors, have their maximum tolerable wattage printed on them. When a resistor carries somewhat more than rated wattage, its resistance begins to change. At still higher wattages, the resistor may burn up.

Other circuit elements, such as most relays, have their maximum rating specified in current rather than watts. This rating system can be converted to watts though, because the resistance of the relay will also be stated, and since

W = EI

and, from Ohm's law,

E = IR

Then

W = IR

Similarly, the wattage rating may sometimes be used to calculate the amount of current flowing through an element. Consider a regular 100-watt household light bulb.

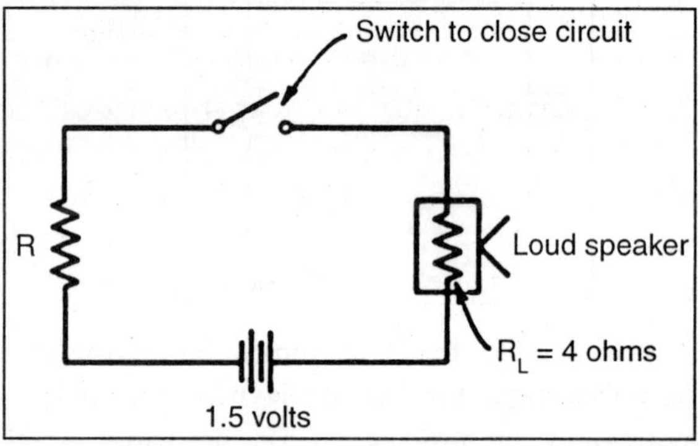

Fig. Circuit to Produce Audible Clicks. Changing the Value of *R* Changes the Loudness of the Clicks.

This 100-watt rating means that, when the light bulb is connected across the normal house voltage, which is typically 115 volts, the light bulb will carry 0.87 ampere.

$W = EI$

$$100 = 115I \; \mathrm{I} = \frac{100}{115} = 0.87 \text{ ampere}$$

Household bulbs are sometimes incompletely labeled. It must be assumed that they are designed for 115 volts in order to make the above calculations. When a device such as a resistor or a flashlight bulb is designed for a maximum voltage other than 115 volts, the voltage is usually specified. Given any two of the four parameters-resistance, voltage, current, watts—the other two may be calculated.

Problems

- A slide projector is used to present stimuli in an experiment. Preliminary trials during which the projector is run from a variable voltage supply (to be explained later) show that the stimulus is at the proper intensity when the voltage across the projector is 60 volts and the current through it is 2 amperes. A variable voltage supply is too expensive to tie up permanently in your experiment. Therefore, determine the resistance and wattage of the resistor which, when put between the 115-volt wall socket and the projector, would cause it to run at the desired intensity.

- You wish to present a subject with a series of clicks at five different loudness levels, the levels being at approximately equal sensory intervals. Such will be the case when the amplitudes of the clicks are in the ratios 1:2:4:8:16 (so long as they are not at either extreme of the loudness range). When a battery is first connected across a loudspeaker, a click is sounded the amplitude of which is approximately proportional to the amount of current flowing through the loudspeaker. Therefore, clicks of different loudnesses can be generated by closing the circuit with different values of R. Assume that you have found, for a loudspeaker whose resistance is 4 ohms, that the loudest click you need occurs when R = 11 ohms. Find the values of the four other resistors that must be substituted for the 11-ohm one in order to produce clicks at the desired loudnesses.

Chapter 2

Transformers

A magnetic field is a construct having to do with the observable fact that objects in the neighborhood of the bar may be affected by the presence of the bar. Figure is a drawing of a bar magnet together with a schematic representation of its magnetic field. Its effect on an object varies depending on where the object is with respect to the bar. This fact is represented by the statement that the strength of the magnetic field varies from place to place around the bar. The lines in Fig. called lines of force, are a schematic representation of the spatial distribution of the strength of the field. The closer the lines, the stronger the field. Since this method of representing a magnetic field facilitates the explanation of many phenomena associated with magnetism, it will be used throughout the chapter.

When a steady current is passed through a coil of wire that is wrapped around an iron core, a magnetic field is set up in and around the core. This magnetic field, represented in Fig, is identical to the field of a permanent bar magnet; the same method of generating magnetism has already been mentioned in connection with meter movements and relays. The magnetic field generated by a steady current is fixed and unchanging even though the electrons ultimately responsible for its maintenance are themselves moving through the coil. However, as soon as the amount of current flowing through the coil is changed in any way, the field changes accordingly.

A transformer is a device that transfers electrical energy from one circuit to another through inductively coupled conductors — the transformer's coils or "windings". Except for air-core transformers, the conductors are commonly wound around a single iron-rich core, or around separate but magnetically-coupled cores. A varying current in the first or "primary" winding creates a varying magnetic field in the core (or cores) of the transformer. This varying magnetic field induces a varying electromotive force (EMF) or "voltage" in the "secondary" winding. This effect is called mutual induction.

If a load is connected to the secondary, an electric current will flow in the secondary winding and electrical energy will flow from the primary circuit through the transformer to the load. In an ideal transformer, the induced voltage in the secondary winding (V_S) is in proportion to the primary voltage (V_P), and is given by the ratio of the number of turns in the secondary to the number of turns in the prima

$$\frac{V_S}{V_P} = \frac{N_S}{N_P}$$

By appropriate selection of the ratio of turns, a transformer thus allows an alternating current (AC) voltage to be "stepped up" by making N_S greater than N_P, or "stepped down" by making N_S less than N_P.

Transformers come in a range of sizes from a thumbnail-sized coupling transformer hidden inside a stage microphone to huge units weighing hundreds of tons used to interconnect portions of national power grids.

All operate with the same basic principles, although the range of designs is wide. While new technologies have eliminated the need for transformers in some electronic circuits, transformers are still found in nearly all electronic devices designed for household ("mains") voltage. Transformers are essential for high voltage power transmission, which makes long distance transmission economically practical.

First Steps: Experiments with Induction Coils

What would become the "transformer principle" was revealed in 1831 by Michael Faraday in his demonstration of electromagnetic induction, but without recognition of its future role in manipulating EMF. The first "induction coils" to see wide use were invented by Rev. Nicholas Callan of Maynooth College, Ireland in 1836, one of the first researchers to realize that the more turns the secondary winding has in relation to the primary winding, the larger the increase in EMF. Induction coils evolved from scientists' and inventors' efforts to get higher voltages from batteries.

Rather than alternating current (AC), their action relied upon a vibrating "make-and-break" mechanism that regularly interrupted the flow of direct current (DC) from the batteries. Between the 1830s and the 1870s, efforts to build better induction coils, mostly by trial and error, slowly revealed the basic principles of transformers. Efficient, practical designs did not appear until the 1880s, but within a decade the "transformer" would be instrumental in the "War of Currents", and in seeing AC distribution systems triumph over their DC counterparts, a position in which they have remained dominant ever since.

In 1876, Russian engineer Pavel Yablochkov invented a lighting system based on a set of induction coils where the primary windings were connected to a source of alternating current and the secondary windings could be connected to several "electric candles" (arc lamps) of his own design. The coils used in the system behaved as primitive transformers. The patent claimed the system could "provide separate supply to several lighting fixtures with different luminous intensities from a single source of electric power".

In 1878, the engineers of the Ganz Company in Hungary assigned part of its extensive engineering works to the manufacture of electric lighting apparatus for Austria-Hungary, and by 1883 made over fifty installations. It offered an entire system consisting of both arc and incandescent lamps, generators, and other accessories. Lucien Gaulard and John Dixon Gibbs first exhibited a device with an open iron core called a "secondary generator" in London in 1882, then sold the idea to the Westinghouse company in the United States. They also exhibited the invention in Turin, Italy in 1884, where it was adopted for an electric lighting system.

Induction coils with open magnetic circuits are inefficient for transfer of power to loads. Various methods of adjusting the cores or bypassing magnetic flux around part of a coil were developed, since until about 1880 the paradigm for AC power transmission from a high voltage supply to a low voltage load was a series circuit.

In practice, several coils with a ratio near 1:1 were connected with their primaries in series to allow use of a high voltage for transmission while presenting a low voltage to the lamps. The inherent flaw in this method was that turning off a single lamp affected all the others on the circuit, and many adjustable coil designs were introduced in an effort to accommodate this problematic characteristic of the series circuit.

Between 1884 and 1885, Hungarian engineers Zipernowsky, Blαthy and Dιri from the Ganz company in Budapest created the efficient "ZBD" closed-core model, which were based on the design by Gaulard and Gibbs. (Gaulard and Gibbs designed just an open core model). They discovered that all former (coreless or open-core) devices were incapable of regulating voltage, and were therefore impracticable.

Their joint patent described a transformer with no poles and comprised two versions of it, the "closed-core transformer" and the "shell-core transformer. In the closed-core transformer the iron core is a closed ring around which the two coils are arranged uniformly. In the shell type transformer, the copper induction cables are passed through the core. In both designs, the magnetic flux linking the primary and secondary coils travels (almost entirely) in the iron core, with no intentional path through air.

The core consists of iron cables or plates. Based on this invention, it became possible to provide economical and cheap lighting for industry and households." Zipernowsky, Blαthy and Dιri discovered the mathematical formula of transformers: Vs/Vp = Ns/Np. With this formula, transformers became calculable and proportionable.

Their patent application made the first use of the word "transformer", a word that had been coined by Ottσ Blαthy. George Westinghouse had bought both Gaulard and Gibbs' and the "ZBD" patents in 1885. He entrusted William Stanley with the building of a ZBD-type transformer for commercial use. Stanley built the core from interlocking E-shaped iron plates. This design was first used commercially in 1886.

Early Developments and Applications

Russian engineer Mikhail Dolivo-Dobrovolsky developed the first three-phase transformer in 1889. In 1891 Nikola Tesla invented the Tesla coil, an air-cored, dual-tuned resonant transformer for generating very high voltages at high frequency. Audio frequency transformers (at the time called repeating coils) were used by the earliest experimenters in the development of the telephone.

Basic Principles

The transformer is based on two principles: firstly, that an electric current can produce a magnetic field (electromagnetism) and secondly that a changing magnetic field within a coil of wire induces a voltage across the ends of the coil (electromagnetic induction). Changing the current in the primary coil changes the magnitude of the applied magnetic field. The changing magnetic flux extends to the secondary coil where a voltage is induced across its ends.

A simplified transformer design is shown to the left. A current passing through the primary coil creates a magnetic field. The primary and secondary coils are wrapped around a core of very high magnetic permeability, such as iron; this ensures that most of the magnetic field lines produced by the primary current are within the iron and pass through the secondary coil as well as the primary coil.

Induction law

The voltage induced across the secondary coil may be calculated from Faraday's law of induction, which states that:

$$V_S = N_S \frac{d\Phi}{dt}$$

where V_S is the instantaneous voltage, N_S is the number of turns in the secondary coil and Φ equals the magnetic flux through one turn of the coil. If the turns of the coil are oriented perpendicular to the magnetic field lines, the flux is the product of the magnetic field strength B and the area A through which it cuts.

The area is constant, being equal to the cross-sectional area of the transformer core, whereas the magnetic field varies with time according to the excitation of the primary. Since the same magnetic flux passes through both the primary and secondary coils in an ideal transformer, the instantaneous voltage across the primary winding equals

$$V_P = N_P \frac{d\Phi}{dt}$$

Taking the ratio of the two equations for V_S and V_P gives the basic equation for stepping up or stepping down the voltage.

Ideal Power Equation

If the secondary coil is attached to a load that allows current to flow, electrical power is transmitted from the primary circuit to the secondary circuit. Ideally, the transformer is perfectly efficient; all the incoming energy is transformed from the primary circuit to the magnetic field and into the secondary circuit. If this condition is met, the incoming electric power must equal the outgoing power.

$$P_{\text{incoming}} = I_P V_P = P_{\text{outgoing}} = I_S V_S$$

giving the ideal transformer equation

If the voltage is increased (stepped up) ($V_S > V_P$), then the current is decreased (stepped down) ($I_S < I_P$) by the same factor. Transformers are efficient so this formula is a reasonable approximation.

The impedance in one circuit is transformed by the *square* of the turns ratio. For example, if an impedance Z_S is attached across the terminals of the secondary coil, it appears to the primary circuit to have an impedance of. This relationship is reciprocal, so that the impedance Z_P of the primary circuit appears to the secondary to be.

Detailed Operation

The simplified description above neglects several practical factors, in particular the primary current required to establish a magnetic field in the core, and the contribution to the field due to current in the secondary circuit.

Models of an ideal transformer typically assume a core of negligible reluctance with two windings of zero resistance. When a voltage is applied to the primary winding, a small current flows, driving flux around the magnetic circuit of the core. The current required to create the flux is termed the *magnetizing current*; since the ideal core has been assumed to have near-zero reluctance, the magnetizing current is negligible, although still required to create the magnetic field.

The changing magnetic field induces an electromotive force (EMF) across each winding. Since the ideal windings have no impedance, they have no associated voltage drop, and so the voltages V_P and V_S measured at the terminals of the transformer, are equal to the corresponding EMFs. The primary EMF, acting as it does in opposition to the primary voltage, is sometimes termed the "back EMF". This is due to Lenz's law which states that the induction of EMF would always be such that it will oppose development of any such change in magnetic field.

Practical Considerations

Leakage Flux

The ideal transformer model assumes that all flux generated by the primary winding links all the turns of every winding, including itself. In practice, some flux traverses paths that take it outside the windings. Such flux is termed *leakage flux*, and results in leakage inductance in series with the mutually coupled transformer windings.

Leakage results in energy being alternately stored in and discharged from the magnetic fields with each cycle of the power supply. It is not directly a power loss, but results in inferior voltage regulation, causing the secondary voltage to fail to be directly proportional to the primary, particularly under heavy load. Transformers are therefore normally designed to have very low leakage inductance. However, in some applications, leakage can be a desirable property, and long magnetic paths, air gaps, or magnetic bypass shunts may be deliberately introduced to a transformer's design to limit the short-circuit current it will supply.

Leaky transformers may be used to supply loads that exhibit negative resistance, such as electric arcs, mercury vapour lamps, and neon signs; or for safely handling loads that become periodically short-circuited such as electric arc welders. Air gaps are also used to keep a transformer from saturating, especially audio-frequency transformers in circuits that have a direct current flowing through the windings.

Effect of Frequency

The time-derivative term in Faraday's Law shows that the flux in the core is the integral of the applied voltage. Hypothetically an ideal transformer would work with direct-current excitation, with the core flux increasing linearly with time. In practice, the flux would rise to the point where magnetic saturation of the core occurs, causing a huge increase in the magnetizing current and overheating the transformer. All practical transformers must therefore operate with alternating (or pulsed) current.

Transformer Universal EMF Equation

The EMF of a transformer at a given flux density increases with frequency. By operating at higher frequencies, transformers can be physically more compact because a given core is able to transfer more power without reaching saturation, and fewer turns are needed to achieve the same impedance. However properties such as core loss and conductor skin effect also increase with frequency. Aircraft and military equipment employ 400 Hz power supplies which reduce core and winding weight.

Operation of a transformer at its designed voltage but at a higher frequency than intended will lead to reduced magnetizing current; at lower frequency, the magnetizing current will increase. Operation of a transformer at other than its design frequency may require assessment of voltages, losses, and cooling to establish if safe operation is practical.

For example, transformers may need to be equipped with "volts per hertz" over-excitation relays to protect the transformer from overvoltage at higher than rated frequency. Knowledge of natural frequencies of transformer windings is of importance for the determination of the transient response of the windings to impulse and switching surge voltages.

Energy Losses

An ideal transformer would have no energy losses, and would be 100% efficient. In practical transformers energy is dissipated in the windings, core, and surrounding structures. Larger transformers are generally more efficient, and those rated for electricity distribution usually perform better than 98%.

Experimental transformers using superconducting windings achieve efficiencies of 99.85%, While the increase in efficiency is small, when applied to large heavily-loaded transformers the annual savings in energy losses are significant. A small transformer, such as a plug-in "wall-wart" or power adapter type used for low-power consumer electronics, may be no more than 85% efficient, with considerable loss even when not supplying any load. Though individual power loss is small, the aggregate losses from the very large number of such devices are coming under increased scrutiny.

The losses vary with load current, and may be expressed as "no-load" or "full-load" loss. Winding resistance dominates load losses, whereas hysteresis and eddy currents losses contribute to over 99% of the no-load loss. The no-load loss can be significant, meaning that even an idle transformer constitutes a drain on an electrical supply, which encourages development of low-loss transformers.

Transformer losses are divided into losses in the windings, termed copper loss, and those in the magnetic circuit, termed iron loss. Losses in the transformer arise from:

Winding Resistance

Current flowing through the windings causes resistive heating of the conductors. At higher frequencies, skin effect and proximity effect create additional winding resistance and losses.

Hysteresis Losses

Each time the magnetic field is reversed, a small amount of energy is lost due to hysteresis within the core. For a given core material, the loss is proportional to the frequency, and is a function of the peak flux density to which it is subjected.

Eddy Currents

Ferromagnetic materials are also good conductors, and a solid core made from such a material also constitutes a single short-circuited turn throughout its entire length. Eddy currents therefore circulate within the core in a plane normal to the flux, and are responsible for resistive heating of the core material. The eddy current loss is a complex function of the square of supply frequency and inverse square of the material thickness.

Magnetostriction

Magnetic flux in a ferromagnetic material, such as the core, causes it to physically expand and contract slightly with each cycle of the magnetic field, an effect known as magnetostriction. This produces the buzzing sound commonly associated with transformers, and in turn causes losses due to frictional heating in susceptible cores.

Mechanical Losses

In addition to magnetostriction, the alternating magnetic field causes fluctuating electromagnetic forces between the primary and secondary windings. These incite vibrations within nearby metalwork, adding to the buzzing noise, and consuming a small amount of power.

Stray Losses

Leakage inductance is by itself lossless, since energy supplied to its magnetic fields is returned to the supply with the next half-cycle. However, any leakage flux that intercepts nearby conductive materials such as the transformer's support structure will give rise to eddy currents and be converted to heat.

Equivalent Circuit

The physical limitations of the practical transformer may be brought together as an equivalent circuit model built around an ideal lossless transformer. Power loss in the windings is current-dependent and is represented as in-series resistances R_P and R_S. Flux leakage results in a fraction of the applied voltage dropped without contributing to the mutual coupling, and thus can be modeled as reactances of each leakage inductance X_P and X_S in series with the perfectly-coupled region.

Iron losses are caused mostly by hysteresis and eddy current effects in the core, and are proportional to the square of the core flux for operation at a given frequency. Since the core flux is proportional to the applied voltage, the iron loss can be represented by a resistance R_C in parallel with the ideal transformer.

A core with finite permeability requires a magnetizing current I_M to maintain the mutual flux in the core. The magnetizing current is in phase with the flux; saturation effects cause the relationship between the two to be non-linear, but for simplicity this effect tends to be ignored in most circuit equivalents.

With a sinusoidal supply, the core flux lags the induced EMF by 90° and this effect can be modeled as a magnetizing reactance (reactance of an effective inductance) X_M in parallel with the core loss component. R_C and X_M are sometimes together termed the *magnetizing branch* of the model. If the secondary winding is made open-circuit, the current I_0 taken by the magnetizing branch represents the transformer's no-load current.

The secondary impedance R_S and X_S is frequently moved (or "referred") to the primary side after multiplying the components by the impedance scaling factor. The resulting model is sometimes termed the "exact equivalent circuit", though it retains a number of approximations, such as an assumption of linearity.

Analysis may be simplified by moving the magnetizing branch to the left of the primary impedance, an implicit assumption that the magnetizing current is low, and then summing primary and referred secondary impedances, resulting in so-called equivalent impedance. The parameters of equivalent circuit of a transformer can be calculated from the results of two transformer tests: open-circuit test and short-circuit test.

Types

A wide variety of transformer designs are used for different applications, though they share several common features. Important common transformer types include:

Autotransformer

An autotransformer has only a single winding with two end terminals, plus a third at an intermediate tap point. The primary voltage is applied across two of the terminals, and the secondary voltage taken from one of these and the third terminal. The primary and secondary circuits therefore have a number of windings turns in common. Since the volts-per-turn is the same in both windings, each develops a voltage in proportion to its number of turns. An adjustable autotransformer is made by exposing part of the winding coils and making the secondary connection through a sliding brush, giving a variable turns ratio.

Polyphase Transformers

For three-phase supplies, a bank of three individual single-phase transformers can be used, or all three phases can be incorporated as a single three-phase transformer. In this case, the magnetic circuits are connected together, the core thus containing a three-phase flow of flux.

A number of winding configurations are possible, giving rise to different attributes and phase shifts. One particular polyphase configuration is the zigzag transformer, used for grounding and in the suppression of harmonic currents.

Leakage Transformers

A leakage transformer, also called a stray-field transformer, has a significantly higher leakage inductance than other transformers, sometimes increased by a magnetic bypass or shunt in its core between primary and secondary, which is sometimes adjustable with a set screw.

This provides a transformer with an inherent current limitation due to the loose coupling between its primary and the secondary windings. The output and input currents are low enough to prevent thermal overload under all load conditions – even if the secondary is shorted.

Leakage transformers are used for arc welding and high voltage discharge lamps (neon lamps and cold cathode fluorescent lamps, which are series-connected up to 7.5 kV AC). It acts then both as a voltage transformer and as a magnetic ballast. Other applications are short-circuit-proof extra-low voltage transformers for toys or doorbell installations.

Resonant Transformers

A resonant transformer is a kind of the leakage transformer. It uses the leakage inductance of its secondary windings in combination with external capacitors, to create one or more resonant circuits. Resonant transformers such as the Tesla coil can generate very high voltages, and are able to provide much higher current than electrostatic high-voltage generation machines such as the Van de Graaff generator.

One of the applications of the resonant transformer is for the CCFL inverter. Another application of the resonant transformer is to couple between stages of a superheterodyne receiver, where the selectivity of the receiver is provided by tuned transformers in the intermediate-frequency amplifiers.

Audio Transformers

Audio transformers are those specifically designed for use in audio circuits. They can be used to block radio frequency interference or the DC component of an audio signal, to split or combine audio signals, or to provide impedance matching between high and low impedance circuits, such as between a high impedance tube (valve) amplifier output and a low impedance loudspeaker, or between a high impedance instrument output and the low impedance input of a mixing console.

Such transformers were originally designed to connect different telephone systems to one another while keeping their respective power supplies isolated, and are still commonly used to interconnect professional audio systems or system components.

Being magnetic devices, audio transformers are susceptible to external magnetic fields such as those generated by AC current-carrying conductors. "Hum" is a term commonly used to describe unwanted signals originating from the "mains" power supply (typically 50 or 60 Hz). Audio transformers used for low-level signals, such as those from microphones, often included shielding to protect against extraneous magnetically-coupled signals.

Instrument Transformers

Instrument transformers are used for measuring voltge, current, power and energy in electrical systems, and for protection and control. Where a voltage or current is too large to be conveniently measured by an instrument, it can be scaled down to a standardized low value. Instrument transformers isolate measurement and control circuitry from the high currents or voltages present on the circuits being measured or controlled.

A current transformer is a transformer designed to provide a current in its secondary coil proportional to the current flowing in its primary coil. Voltage transformers (VTs), also referred to as "potential transformers" (PTs), are used in high-voltage circuits. They are designed to present a negligible load to the supply being measured, to allow protective relay equipment to be operated at lower voltages, and to have a precise winding ratio for accurate metering.

Classification

Transformers can be classified in different ways:

- *By power capacity*: from a fraction of a volt-ampere (VA) to over a thousand MVA;
- *By frequency range*: power-, audio-, or radio frequency;
- *By voltage class*: from a few volts to hundreds of kilovolts;
- *By cooling type*: air cooled, oil filled, fan cooled, or water cooled;

- *By application*: such as power supply, impedance matching, output voltage and current stabilizer, or circuit isolation;
- *By end purpose*: distribution, rectifier, arc furnace, amplifier output;
- *By winding turns ratio*: step-up, step-down, isolating (equal or near-equal ratio), variable.

Construction

Cores

Laminated Steel Cores

Transformers for use at power or audio frequencies typically have cores made of high permeability silicon steel. The steel has a permeability many times that of free space, and the core thus serves to greatly reduce the magnetizing current, and confine the flux to a path which closely couples the windings.

Early transformer developers soon realized that cores constructed from solid iron resulted in prohibitive eddy-current losses, and their designs mitigated this effect with cores consisting of bundles of insulated iron wires. Later designs constructed the core by stacking layers of thin steel laminations, a principle that has remained in use. Each lamination is insulated from its neighbors by a thin non-conducting layer of insulation. The universal transformer equation indicates a minimum cross-sectional area for the core to avoid saturation.

The effect of laminations is to confine eddy currents to highly elliptical paths that enclose little flux, and so reduce their magnitude. Thinner laminations reduce losses, but are more laborious and expensive to construct. Thin laminations are generally used on high frequency transformers, with some types of very thin steel laminations able to operate up to 10 kHz.

One common design of laminated core is made from interleaved stacks of E-shaped steel sheets capped with I-shaped pieces, leading to its name of "E-I transformer". Such a design tends to exhibit more losses, but is very economical to manufacture. The cut-core or C-core type is made by winding a steel strip around a rectangular form and then bonding the layers together. It is then cut in two, forming two C shapes, and the core assembled by binding the two C halves together with a steel strap. They have the advantage that the flux is always oriented parallel to the metal grains, reducing reluctance.

A steel core's remanence means that it retains a static magnetic field when power is removed. When power is then reapplied, the residual field will cause a high inrush current until the effect of the remaining magnetism is reduced, usually after a few cycles of the applied alternating current. Overcurrent protection devices such as fuses must be selected to allow this harmless inrush to pass. On transformers connected to long, overhead power transmission lines, induced currents due to geomagnetic disturbances during solar storms can cause saturation of the core and operation of transformer protection devices.

Distribution transformers can achieve low no-load losses by using cores made with low-loss high-permeability silicon steel or amorphous (non-crystalline) metal alloy. The higher initial cost of the core material is offset over the life of the transformer by its lower losses at light load.

Solid Cores

Powdered iron cores are used in circuits (such as switch-mode power supplies) that operate above main frequencies and up to a few tens of kilohertz. These materials combine high magnetic permeability with high bulk electrical resistivity. For frequencies extending beyond the VHF band, cores made from non-conductive magnetic ceramic materials called ferrites are common. Some radio-frequency transformers also have movable cores (sometimes called 'slugs') which allow adjustment of the coupling coefficient (and bandwidth) of tuned radio-frequency circuits.

Toroidal Cores

Toroidal transformers are built around a ring-shaped core, which, depending on operating frequency, is made from a long strip of silicon steel or permalloy wound into a coil, powdered iron, or ferrite. A strip construction ensures that the grain boundaries are optimally aligned, improving the transformer's efficiency by reducing the core's reluctance.

The closed ring shape eliminates air gaps inherent in the construction of an E-I core. The cross-section of the ring is usually square or rectangular, but more expensive cores with circular cross-sections are also available. The primary and secondary coils are often wound concentrically to cover the entire surface of the core. This minimizes the length of wire needed, and also provides screening to minimize the core's magnetic field from generating electromagnetic interference.

Toroidal transformers are more efficient than the cheaper laminated E-I types for a similar power level. Other advantages compared to E-I types, include smaller size (about half), lower weight (about half), less mechanical hum (making them superior in audio amplifiers), lower exterior magnetic field (about one tenth), low off-load losses (making them more efficient in standby circuits), single-bolt mounting, and greater choice of shapes.

The main disadvantages are higher cost and limited power capacity.

Ferrite toroidal cores are used at higher frequencies, typically between a few tens of kilohertz to a megahertz, to reduce losses, physical size, and weight of switch-mode power supplies.

A drawback of toroidal transformer construction is the higher cost of windings. As a consequence, toroidal transformers are uncommon above ratings of a few kVA. Small distribution transformers may achieve some of the benefits of a toroidal core by splitting it and forcing it open, then inserting a bobbin containing primary and secondary windings.

Air Cores

A physical core is not an absolute requisite and a functioning transformer can be produced simply by placing the windings in close proximity to each other, an arrangement termed an "air-core" transformer. The air which comprises the magnetic circuit is essentially lossless, and so an air-core transformer eliminates loss due to hysteresis in the core material.

The leakage inductance is inevitably high, resulting in very poor regulation, and so such designs are unsuitable for use in power distribution. They have however very high bandwidth, and are frequently employed in radio-frequency applications, for which a satisfactory coupling coefficient is maintained by carefully overlapping the primary and secondary windings.

Windings

The conducting material used for the windings depends upon the application, but in all cases the individual turns must be electrically insulated from each other to ensure that the current travels throughout every turn. For small power and signal transformers, in which currents are low and the potential difference between adjacent turns is small, the coils are often wound from enameled magnet wire, such as Formvar wire. Larger power transformers operating at high voltages may be wound with copper rectangular strip conductors insulated by oil-impregnated paper and blocks of pressboard.

High-frequency transformers operating in the tens to hundreds of kilohertz often have windings made of braided Litz wire to minimize the skin-effect and proximity effect losses. Large power transformers use multiple-stranded conductors as well, since even at low power frequencies non-uniform distribution of current would otherwise exist in high-current windings.

Each strand is individually insulated, and the strands are arranged so that at certain points in the winding, or throughout the whole winding, each portion occupies different relative positions in the complete conductor. The transposition equalizes the current flowing in each strand of the conductor, and reduces eddy current losses in the winding itself. The stranded conductor is also more flexible than a solid conductor of similar size, aiding manufacture.

For signal transformers, the windings may be arranged in a way to minimize leakage inductance and stray capacitance to improve high-frequency response. This can be done by splitting up each coil into sections, and those sections placed in layers between the sections of the other winding. This is known as a stacked type or interleaved winding.

Both the primary and secondary windings on power transformers may have external connections, called taps, to intermediate points on the winding to allow selection of the voltage ratio. The taps may be connected to an automatic on-load tap changer for voltage regulation of distribution circuits. Audio-frequency transformers, used for the distribution of audio to public address loudspeakers, have taps to allow adjustment of impedance to each speaker.

A centre-tapped transformer is often used in the output stage of an audio power amplifier in a push-pull circuit. Modulation transformers in AM transmitters are very similar.

Certain transformers have the windings protected by epoxy resin. By impregnating the transformer with epoxy under a vacuum, one can replace air spaces within the windings with epoxy, thus sealing the windings and helping to prevent the possible formation of corona and absorption of dirt or water. This produces transformers more suited to damp or dirty environments, but at increased manufacturing cost.

Coolant

High temperatures will damage the winding insulation. Small transformers do not generate significant heat and are cooled by air circulation and radiation of heat. Power transformers rated up to several hundred kVA can be adequately cooled by natural convective air-cooling, sometimes assisted by fans.

In larger transformers, part of the design problem is removal of heat. Some power transformers are immersed in transformer oil that both cools and insulates the windings. The oil is a highly refined mineral oil that remains stable at transformer operating temperature. Indoor liquid-filled transformers must use a non-flammable liquid, or must be located in fire resistant rooms. Air-cooled dry transformers are preferred for indoor applications even at capacity ratings where oil-cooled construction would be more economical, because their cost is offset by the reduced building construction cost.

The oil-filled tank often has radiators through which the oil circulates by natural convection; some large transformers employ forced circulation of the oil by electric pumps, aided by external fans or water-cooled heat exchangers. Oil-filled transformers undergo prolonged drying processes to ensure that the transformer is completely free of water vapour before the cooling oil is introduced. This helps prevent electrical breakdown under load. Oil-filled transformers may be equipped with Buchholz relays, which detect gas evolved during internal arcing and rapidly de-energize the transformer to avert catastrophic failure.

Polychlorinated biphenyls have properties that once favored their use as a coolant, though concerns over their environmental persistence led to a widespread ban on their use. Today, non-toxic, stable silicone-based oils, or fluorinated hydrocarbons may be used where the expense of a fire-resistant liquid offsets additional building cost for a transformer vault. Before 1977, even transformers that were nominally filled only with mineral oils may also have been contaminated with polychlorinated biphenyls at 10-20 ppm. Since mineral oil and PCB fluid mix, maintenance equipment used for both PCB and oil-filled transformers could carry over small amounts of PCB, contaminating oil-filled transformers.

Some "dry" transformers (containing no liquid) are enclosed in sealed, pressurized tanks and cooled by nitrogen or sulfur hexafluoride gas. Experimental power transformers in the 2 MVA range have been built with superconducting windings which eliminates the copper losses, but not the core steel loss. These are cooled by liquid nitrogen or helium.

Terminals

Very small transformers will have wire leads connected directly to the ends of the coils, and brought out to the base of the unit for circuit connections. Larger transformers may have heavy bolted terminals, bus bars or high-voltage insulated bushings made of polymers or porcelain. A large bushing can be a complex structure since it must provide careful control of the electric field gradient without letting the transformer leak oil.

Applications

A major application of transformers is to increase voltage before transmitting electrical energy over long distances through wires. Wires have resistance and so dissipate electrical energy at a rate proportional to the square of the current through the wire. By transforming electrical power to a high-voltage (and therefore low-current) form for transmission and back again afterward, transformers enable economic transmission of power over long distances. Consequently, transformers have shaped the electricity supply industry, permitting generation to be located remotely from points of demand. All but a tiny fraction of the world's electrical power has passed through a series of transformers by the time it reaches the consumer.

Transformers are also used extensively in electronic products to step down the supply voltage to a level suitable for the low voltage circuits they contain. The transformer also electrically isolates the end user from contact with the supply voltage.

Signal and audio transformers are used to couple stages of amplifiers and to match devices such as microphones and record player s to the input of amplifiers. Audio transformers allowed telephone circuits to carry on a two-way conversation over a single pair of wires. Transformers are also used when it is necessary to couple a differential-mode signal to a ground-referenced signal, and for between external cables and internal circuits.

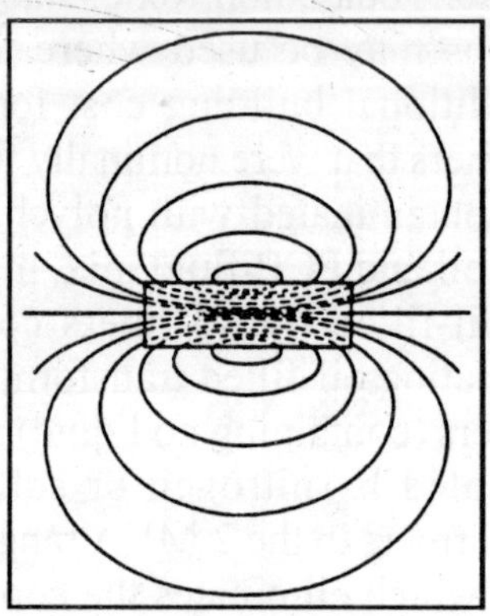

Fig. Schematic Representation of a Bar Magnet and its Magnetic Field. All the Lines are Actually Loops, but some of the Loops are too Large to be Represented on the Diagram.

For instance, if the current should suddenly be doubled, the field would double in strength, as represented in Fig. After doubling the current, there will be twice as many lines of force spread out over a larger volume than before. If one of these hypothetical lines of force were visible while the current were in the process of doubling, it would be seen to move outward.

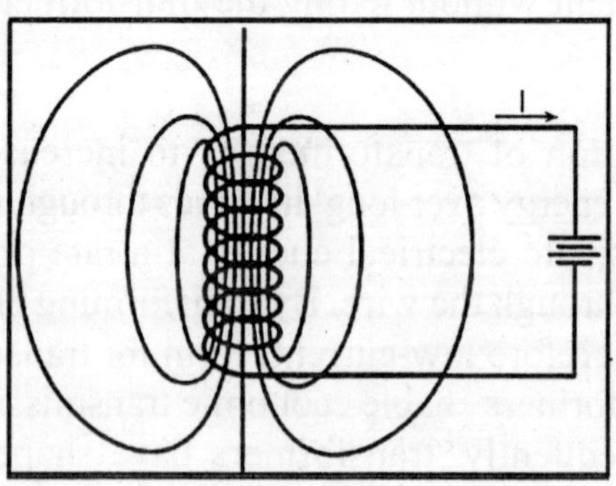

Fig. Schematic Representation of the Magnetic Field Generated by the Passage of Current I Through a Coil of Wire.

In other words, increasing the current through the coil not only adds more lines to the field but also causes the field to expand. Conversely, if the current were reduced, the field would collapse. Now store that concept for a moment and consider the opposite sort of phenomenon, namely, the effect that a magnetic field has on a coil of wire. If you have a permanent magnet at hand, you can demonstrate certain phenomena very readily. Connect your meter directly across the two ends of the coil on your sensitive relay, and set the meter to its most sensitive scale (e.g., 100 microamperes, 1 volt, etc.).

Then put the magnet down next to the relay coils. Some of the lines of force from the magnet will now be passing through the relay coil but you will notice that nothing happens to the meter. In general, a stationary magnetic field has no effect on a coil of wire. But a moving field does affect the coil. Any time that lines of force move with respect to a coil of wire, they induce a voltage into it. (There is one particular direction of motion of a field with respect to a coil which will not induce a voltage into the coil, but that special case will be ignored for the present discussion.) It is convenient to think of this process as one in which each line of magnetic force tends to push some of the free electrons in the wire along in front of it as it moves.

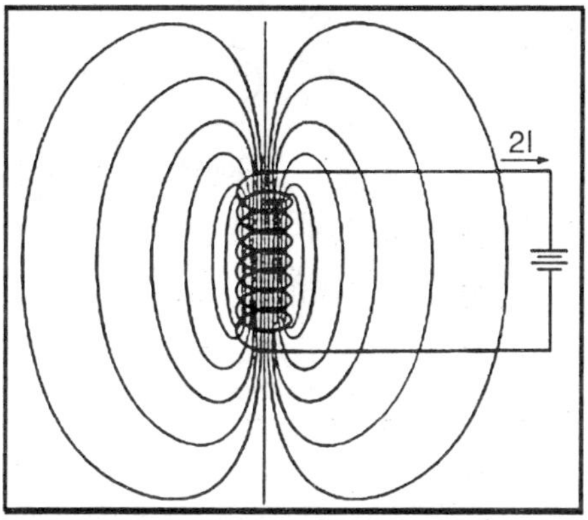

Fig. The Magnetic Field of the Coil when the Current Through it is Doubled. The Field is Represented as having Twice as many Lines, Distributed Over a Larger Area (Volume).

You can observe this by moving the magnet rapidly past the relay coil so that its lines of force move through the coil. If you have a reasonably strong magnet and if the meter is set at a sensitive scale, the pointer will jump to the right when the magnet passes in one direction, and to the left when it moves in the opposite direction. The lines of force are pushing electrons one way or the other. Note, though, that the needle reads nonzero only when the field is actually moving.

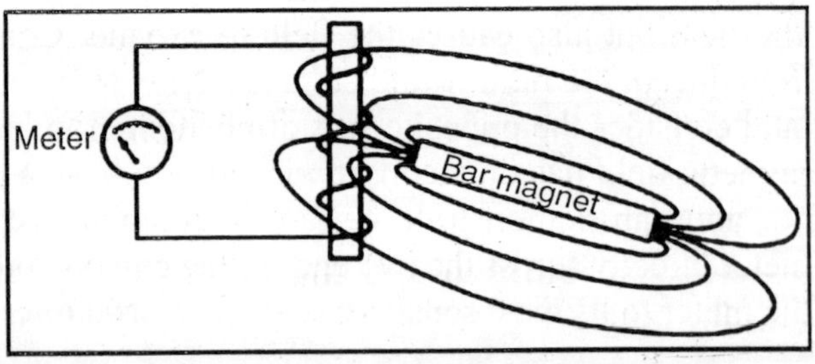

Fig. Representation of a Bar Magnet Close to a Relay Coil. Some of the Lines of Force Intersect the Coil. So Long as the Magnet is Stationary with Respect to the Coil, no Current will Flow Around the Loop Containing the Meter.

These two phenomena, the generation of a magnetic field by current flowing in a coil and the induction of current in a coil by a moving magnetic field, are combined in an extensively used device called the transformer. Its principle of operation is fairly simple. The switch is open and no current flows in either circuit.

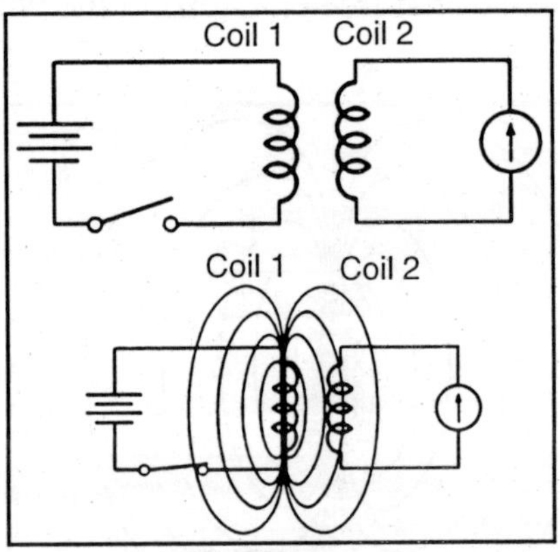

Fig. (*a*) A Circuit Containing two Coils of Wire, Near Each Other but not Connected Together Electrically. (*b*) Representation of the Circuit after the Switch has been Closed. Some of the Lines of Force from Coil I Intersect Coil 2, but so Long as the Current Through Coil I is Constant, no Current will Flow Through Coil 2.

Figure *b* shows the state of affairs when the switch has been closed and current is flowing steadily through coil 1. Although the lines of force from coil 1 are passing through coil 2, no voltage is generated in coil 2 because the lines are not moving. However, between the time when no current flows through coil 1 and when the full current is flowing, that is the lines of force must have multiplied and expanded through coil 2 from zero lines and volume to their positions. During the time when the field was being built up, a voltage must have been induced into coil 2. Similarly, if the switch were opened again, the field would collapse and, during that short time, a voltage would be induced into coil 2 as well. The voltage in coil 2 would have the opposite polarity in the two cases because the lines of force would be traveling in opposite directions during the buildup and collapse of the field. The buildup and the collapse do not occur instantaneously but take finite times which depend upon the characteristics of the circuit. Now suppose the battery and switch were replaced by an a-c source, for instance, the 115-volt a-c wall socket supply. In this case, the current through coil 1 will build up from zero to a maximum, then back through zero again, etc., repeating this complete cycle 60 times a second.

In turn, the magnetic field will also build up and collapse over and over again, continuously generating a changing and reversing voltage in coil 2. This kind of combination of coils is called a transformer. For more efficient transfer of power from coil I to coil 2, both coils are typically wound on a single iron core, because such a core tends to channel the magnetic fields, keeping the lines of force in the close vicinity of the coils. The schematic diagram for a transformer, together with photographs of actual transformers.

If coil 1 consists of the same number of turns of wire as coil 2, the voltage generated in coil 2 will be the same as that put into coil 1. However, if there are, say, ten times as many turns on coil 2 as on coil 1, the lines of force from coil 1 will cut across coil 2 at ten times as many places.

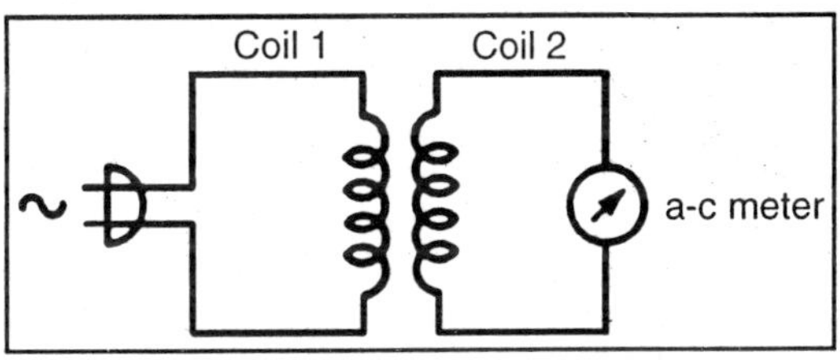

Fig. Diagram of a Simple Transformer Circuit. The Meter will Indicate a Continuous Voltage.

Therefore the voltage generated in coil 2 will be ten times as great as the input voltage to coil 1. In general:

$$\frac{V_i}{V_o} = \frac{N_i}{N_o}$$

where *Vi* = the input voltage *Vo* = the output voltage *Ni* = the number of turns on the input coil *No* = the number of turns on the output coil

It is this phenomenon that makes the transformer so useful. For example, suppose that you were designing a circuit to give a rat a strong constant-current shock. For the shock to have a relatively constant current, a large resistance must be in series with the rat. In fact, the resistance must be large enough that almost all of the voltage from the source is lost across the resistor, and only about one-tenth of it actually affects the rat.

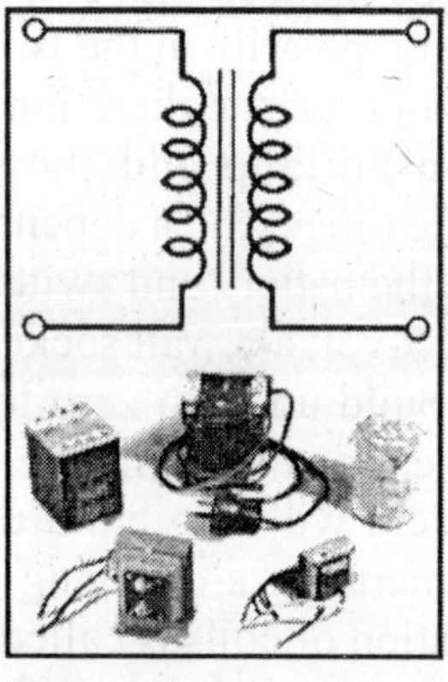

Fig. (*a*) Circuit Symbol for a Transformer, and (*b*) a Photograph of Various Transformers.

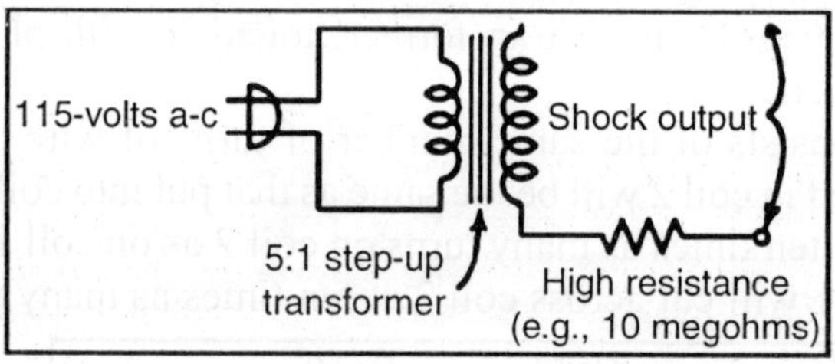

Fig. Constant-current a-c Shock-Delivery Circuit.

To give the rat a good strong shock, the source voltage must be higher than the 115 volts from the wall. A 50-volt shock, for example, requires a 500-volt source. The solution is to put between the wall socket and the shock box a transformer which has about five times as many turns on the output (or secondary) coil as it has on the input (or primary) coil. The transformer is called a step-up transformer.

Electric-eye systems which detect the presence of a rat in an alley require a small source of light such as a flashlight bulb or an automobile taillight bulb. These bulbs are typically rated at 6 volts.

Flashlight batteries could be used to light the bulbs, but they burn out quickly, and storage batteries are inconvenient. However, a small *step-down* transformer which takes the 115 volts from a wall socket and transforms it to 6 volts is cheap, available, and never wears out. Such a transformer has approximately 20 times as many turns on its input coil as on its output coil.

Note that a transformer is so named because it transforms one alternating voltage into another alternating voltage. It does not transform direct current into alternating current (that is accomplished by a device called a converter) nor alternating current into direct current (that device is called a rectifier). A transformer is strictly an a-c device, and as such can be quickly and irreversibly damaged if it is connected across a d-c source.

TRANSFORMER RATINGS

When buying a transformer you should specify:

- Input (primary) voltage and frequency (usually 115 volts, 60 cycles).
- Output (secondary) voltage. Output (secondary) current.
- Other special requirements such as size, etc.

INPUT (PRIMARY) VOLTAGE

The transformer is most commonly used to provide a convenient source of a-c power at a voltage other than the one provided at the wall socket. There is only one type of transformer used at all frequently in the laboratory that has a rated input voltage or frequency other than 115 volts, 60 cycles. It is called an output transformer and is used in certain electronic circuits which will not be discussed here.

Many transformers found in surplus stores are rated for 400-cycle voltages. They are common because the standard frequency for certain military devices, such as airplanes, is 400 cycles rather than 60 cycles. A 400-cycle transformer will not operate properly on a 60cycle voltage.

OUTPUT (SECONDARY) VOLTAGE

The output, or secondary voltage, is simply the voltage that the transformer will deliver across its secondary coil terminals when the rated input voltage is applied to its primary coil. If you wish to light a 6-volt bulb from a wall socket, order a transformer with an output voltage of 6 volts. Many transformers are built to give a variety of output voltages all at the same time.

This is sometimes accomplished by winding two or more electrically separate output coils on the same core. That arrangement is very common in transformers designed to supply power for electronic devices. Most radios, for instance, require a low voltage, e.g., 6 volts, to light the tubes, and a high voltage, e.g., 300 volts, to drive electrons through the vacuum in the tubes.

Multiple-output voltages are also commonly obtained by the method. Connections are made at different places along a single output coil. It has already been pointed out that the more turns there are on the secondary winding, the higher the voltage that is generated.

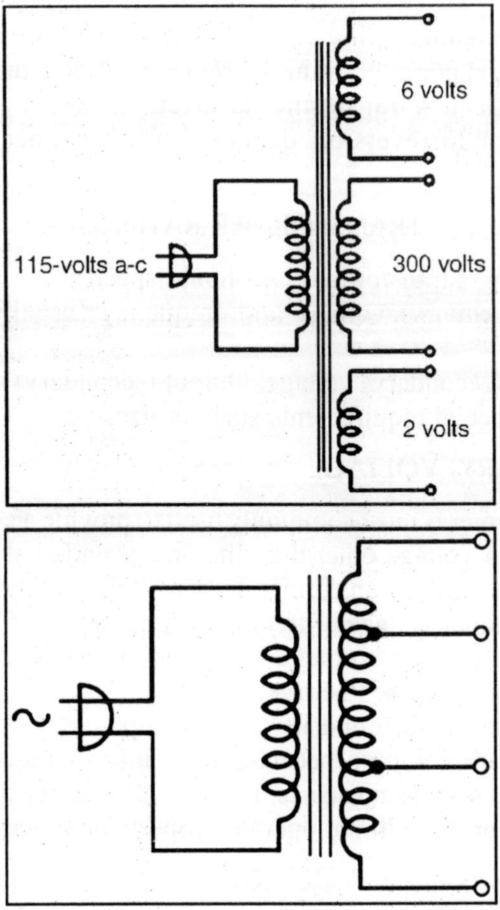

Fig. (*a*) Diagram of a Transformer with Three Different Output Windings.
(*b*) Diagram of a Transformer with a Multiple-tapped Output Winding.

It should be evident that different pairs of terminals have different numbers of turns between them and will, therefore, deliver different voltages. The wires connected along the coil are called taps, and this kind of transformer is said to have a multiple-tapped output winding. The most common tap is one halfway between the two ends of the output coil and such a transformer is said to be centre-tapped. Thus, if you want a transformer that will deliver either 6 or 12 volts, ask for a 12-volt-output, centre-tapped transformer.

OUTPUT CURRENT

Suppose that you wish to light three 6-volt, 5-ampere bulbs in parallel. You must order a transformer that gives a 6-volt output with an output current rating of at least 15 amperes; that is, the output coil must be able to carry 15 amperes without getting too hot. Note that this is a maximum current rating. If a transformer has a 6-volt, 15ampere output rating, it will always put out about 6 volts, but it will not necessarily deliver 15 amperes at all times. If there is nothing connected between the two terminals of the output coil, it will not deliver any current at all (6 volts/infinite ohms = 0 ampere). But if there is less than 0.4 ohm between the output terminals, more than the rated current will flow, and the transformer will get too hot.

CIRCUITS USING TRANSFORMERS

The circuit will light three automobile headlight bulbs in parallel. If these bulbs draw 5 amperes each, the current labeled I2 will, of course, be 15 amperes a-c. But the value of I1 is not so obvious. For any transformer the ratio of the input to output current is the inverse of the ratio of the input to output voltage. If the input voltage of the transformer is twice the output voltage, the input current will always be half the output current. This is expressed as

$$\frac{V_{\text{input}}}{V_{\text{output}}} = \frac{I_{\text{output}}}{I_{\text{input}}}$$

Note that this equation can also be expressed as

Vinput X Iinput = Voutput X Ioutput

(i.e.)

Input watts = output watts

In other words, if a certain amount of power is coming out of a transformer, the same amount will be going in. This makes sense as long as the transformer itself does not use up much power. Most transformers operate so efficiently that errors introduced by using these equations are negligible. In the example, we are given the output current (15 amperes), the output voltage (6 volts), and the input voltage (115 volts). Therefore

$$I_{\text{input}} = \frac{15 \times 6}{115} = 0.78 \text{ ampere}$$

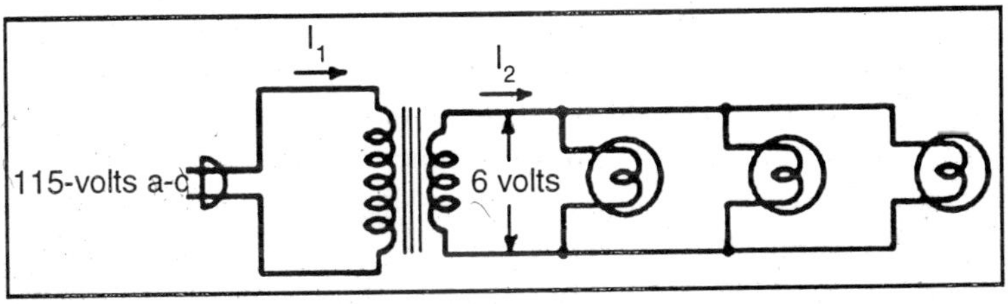

Fig. Circuit to Light Three Automobile Headlight Bulbs in Parallel. I_1, is the Input Current and I_2 t he Output Current of the Transformer.

To extend this treatment, consider what happens when there is no connection (an infinite resistance) between the two output terminals of the transformer. The output voltage will still be 6 volts, but now the output current is zero. The output power is zero watts, and therefore the input power must also be zero watts. Since the input voltage is not zero, the input current must be zero. In other words, according to this equation, a transformer plugged into the wall will not draw any current in its primary as long as nothing is connected across its secondary. This conclusion looks peculiar but is true for all practical purposes.

ESTIMATING THE RATINGS

Transformers which have no ratings engraved on them can usually be found in the dark corners of the laboratory. This section will describe several ways in which you may start to classify them. First of all, if what you think might be a transformer has only two wires coming out of it, it is a choke rather than a transformer. A choke looks like a transformer but it consists of only a single coil of wire on an iron core. If there are either three of four terminals on the unknown transformer and they are labeled with letters such as *b*, *p*, or *g*, it is an output transformer and should not be used except in certain electronic applications which will not be discussed here. But if the transformer has at least four terminals or wires, and if they are not labeled with the letters, *b*, *p*, or *g*, it may be usable.

The easiest way to determine its output voltage is to apply 115 volts, 60 cycles per second, to the primary coil and measure the voltages resulting in the secondary or secondaries. On some transformers the primary is labeled "primary," or "input," or "115 volts a-c," and then this procedure is straightforward, but usually there are no such markings and you have to guess which wires connect to the input coil. When there are only four wires or terminals and none is labeled, the four must be divided into two pairs, one pair from the input and the other from the output coil. If the pairing is not obvious from the geography of the transformer, it can be determined simply by using an ohmmeter to find out which pairs of wires are connected together. Now examine the leads themselves.

If one pair consists of fairly heavy wires, it is very likely that the unit is a step-down transformer and that the heavy pair are the secondary leads. If one pair of terminals is mounted on a large ceramic holder, that pair is probably the output of a very-high-voltage transformer (e.g., 5000 volts or more). If neither is true, either coil can be used as the primary and the other as the secondary. In one case you will have a step-up transformer, and in the other a step-down transformer. To find out the ratio of input to output voltage, one pair of terminals may be connected across the 115-volt wall voltage, and the voltage between the others measured with an a-c voltmeter. However, since it is possible that the application of 115 volts to one of the coils will damage it, a safer procedure is to increase the input voltage gradually from zero to 115 volts, stopping if the transformer begins to heat up.

If there are five wires on an unknown transformer, it is probable that the fifth wire is a centre tap on the output coil. If so, an ohmmeter will indicate that there is one pair and one trio of connected wires. The output voltages (two of them) can then be measured by applying a known a-c voltage across the pair and measuring the voltages between pairs of the three remaining wires. When there are more than five wires or terminals on a transformer, the job becomes a lot more difficult. In general, the best procedure is to find out with an ohmmeter which wires are connected to each other, and then try to reconstruct the circuit diagram of the transformer from this information.

Suppose you find a transformer with eight wires coming out of its casing, and an ohmmeter indicates that there are one pair and two groups of three. Because primaries rarely have centre taps, the pair is very likely the primary. Then examine the wires themselves. If in one of the groups of three, two of the wires are thick, the chances are very good that the group of three is a low-voltage output coil with a centre tap (probably 6 volts), and that the other group of three is a high-voltage output coil with a centre tap (e.g., 300 volts).

It is very difficult for an unpracticed worker to judge the output current rating of an unknown transformer. The only real way to determine this is to let the transformer draw some current and see if it gets hot. This is done by putting the rated voltage across the input terminals and connecting a variable resistor across the output. Then gradually increase the current by decreasing the resistance until the transformer either gets hot or is delivering the required current. A good criterion for "hot" is this: If you can hold your hand against the transformer, it is not overheating. If you have to pull your hand away, the transformer is overloaded. Any transformer running near its rated load will be hot, but it should not be so hot that you cannot hold your hand against it.

ADJUSTABLE-VOLTAGE TRANSFORMERS

One of the most useful gadgets around a laboratory is the adjustable voltage transformer, often referred to as "Variac." (Variac is a trade name for a particular, widely sold brand of adjustable voltage transformer.) The customary input voltage to this type of transformer is 115 volts, 60 cycles, and the output can be adjusted continuously from zero to 115 or zero to 140 volts by turning a knob on the top.

The theory of operation of this sort of transformer is too complicated to be discussed herein. At any given setting of the slider arm, it acts like a normal, 2-coil transformer except that there is a direct electric connection between the input and the output. This connection is sometimes important. Adjustable voltage transformers are very convenient for changing the intensity of light bulbs, the speed of motors, etc.

They are particularly useful for testing devices the ratings of which are unknown. For example, suppose that you need a relay and you find one in the stockpile, but the coil voltage required to make it operate is unknown.

Connect it across the output of an adjustable-voltage transformer and gradually increase the voltage from zero until it closes firmly. Then read the required voltage with a voltmeter. (If the relay never closes firmly but just buzzes, it is a d-c relay.) The same procedure can be applied to other devices such as unlabeled motors or transformers: increase the applied voltage gradually until the devices operate properly.

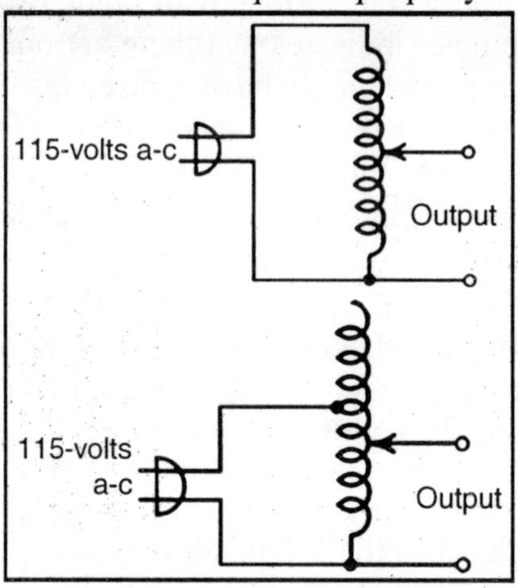

Fig. (*a*) Diagram of a Transformer whose output Voltage is Adjustable from 0 to 115 Volts. (*b*) Diagram of a 0 to 140-volt Adjustable Voltage Transformer.

ISOLATION TRANSFORMERS

There is a class of transformers called isolation transformers which usually have the same number of turns of wire on the primary as on the secondary, so that they neither step up nor step down the input voltage.

They are used when it is desired to isolate a circuit electrically from the source of power. For example, suppose that you constructed the very simple device to deliver a weak electric shock to a human subject. One side of the wall socket is already connected to "ground," that is, to the earth itself, and to the water pipes, radiator, etc. This side of the socket is so labeled.

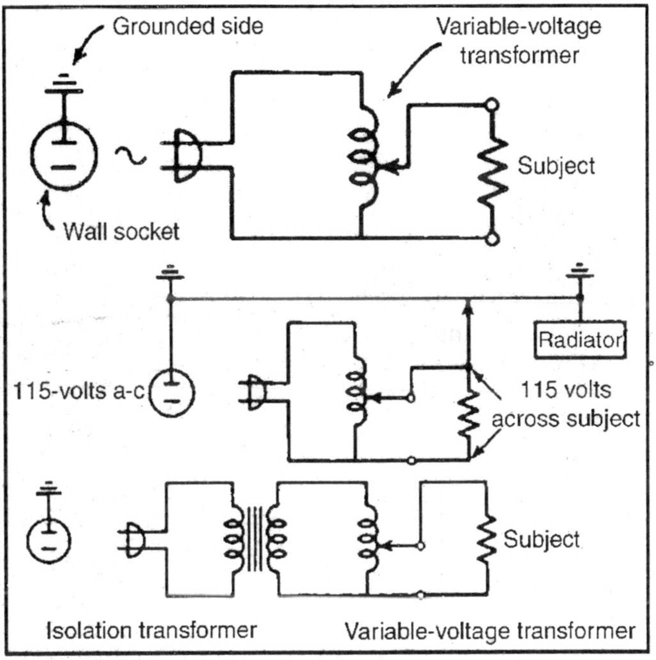

Fig. (*a*) Circuit to Shock Subjects. This Circuit Should not be Used Because of the Serious Shock Hazard. (*b*) Diagram of the Actual Circuit when a Subject, being Shocked by the Circuit Touches any Grounded Object. The Full wall-socket Voltage Appears Across the Subject. (*c*) A Safe Circuit for Shocking Human Subjects. The Isolation Transformer Eliminates the Shock Hazard of the Circuit.

Now suppose that the subject reaches over and touches some metal object that is in contact with a radiator. He will be completing the circuit, and will no longer be a useful subject. This possibility can be eliminated by placing an isolation transformer between the adjustable voltage transformer and the wall socket.

CONSTANT VOLTAGE TRANSFORMERS

The voltage at the wall socket is usually said to be 115 volts. Actually, that voltage varies widely from time to time. It is a common experience to see the room lights dim when the refrigerator goes on. The refrigerator motor, when it first turns on, draws a large current which flows through the wires leading into the house and to the wall socket. The increased current running through the resistance of those wires causes an increased voltage drop in the wires, and less Voltage is left to hear the lights.

In and around the laboratory, and all along the circuit that includes the power station and the laboratory, devices turn on and off repeatedly, so that it is not unusual for the voltage at the wall to vary from 90 to 130 volts during the day. (The voltage can be higher than 115 because, to overcome the average voltage loss in the wires, the power company delivers more than 115 volts to the system. If everyone but you should suddenly turn off everything electrical, all of your light bulbs would burn out.)

For many devices, such changes in voltage do not matter very much (e.g., fan motors, relays), but several types of apparatus are very sensitive to voltage changes and must be protected from them. All laboratory apparatus in which the intensity of a light bulb is important fall into this category because relatively small changes in line voltage will produce large changes in any lamp's brightness. For example, variations in the line voltage can be a source of great variance in tachistoscopic studies.

Devices which take variable a-c inputs and deliver constant-voltage a-c outputs are called voltage regulators or voltage stabilizers and may be ordered at any electronics supply store. There are a number of different types of voltage regulators, some considerably more costly than others. The cheapest, and one which meets most requirements, is called a constant-voltage transformer. It has no moving parts and is essentially a standard transformer built around a special core. The molecules of metal in any transformer core are involved in the transfer of energy from the input to the output coil. In a constant-voltage transformer, all of the core molecules are brought into action when the instantaneous input voltage is, say, 90 volts. The coils are wound so that the output voltage is 115 volts under this condition.

Any further increase in the input voltage cannot affect the output voltage appreciably because all of the core molecules are already involved (the core is said to be saturated). This transformer will, therefore, give almost constant output voltage for input voltages ranging from 90 to some higher voltage where the system begins to break down. Typically, the upper limit is approximately 130 volts. Transformers of this type are built with various output-current capacities from about 1/10 ampere up to 10 or 20 amperes. The rating is usually stated in units, called volt-amperes, which are very closely related to watts. If you want 115 volts at about 4 amperes, buy a 500-volt-ampere unit. (For a-c circuits, power—in watts—is sometimes different from the simple product of volts times amperes.)

The constant-voltage transformers described above have several defects which may be of consequence in certain applications. First, they only hold the voltage stable to about 11% of its nominal level. For the rare situation in which something better is needed, a more expensive unit must be bought. Second, these transformers usually hum and, because of this, should not be used when absolute quiet is needed. A third defect is that, since the core saturates in the normal range of input voltages, the output wave form is a flat-topped wave that approaches the shape of a square wave. For the rare occasions when a very low harmonic content is required of the supply voltage, regulators which give pure sine-wave outputs can be purchased.

Voltage regulators are not to be confused with the very common piece of electronic equipment called a regulated power supply. The term "power supply," when it is applied to a single unit of equipment, usually refers to a unit that supplies d-c power. A regulated, or stabilized, power supply unit is one that delivers a constant d-c output voltage independent of changes in the line voltage.

TRANSFORMER TYPES

A variety of types of electrical transformer are made for different purposes. Despite their design differences, the various types employ the same basic principle as discovered in 1831 by Michael Faraday, and share several key functional parts.

Power transformers

Laminated core

This is the most common type of transformer, widely used in appliances to convert mains voltage to low voltage to power electronics

- Widely available in power ratings ranging from mW to MW
- Insulated laminations minimize eddy current losses
- Small appliance and electronic transformers may use a split bobbin, giving a high level of insulation between the windings
- Rectangular core
- Core laminate stampings are usually in EI shape pairs. Other shape pairs are sometimes used.
- Mumetal shields can be fitted to reduce EMI (electromagnetic interference)
- A screen winding is occasionally used between the 2 power windings
- Small appliance and electronics transformers may have a thermal cut out built in
- Occasionally seen in low profile format for use in restricted spaces
- Laminated core made with silicon steel with high permeability

Toroidal

Doughnut shaped toroidal transformers are used to save space compared to EI cores, and sometimes to reduce external magnetic field. These use a ring shaped core, copper windings wrapped round this ring (and thus threaded through the ring during winding), and tape for insulation.

Toroidals compared to EI core transformers:

- Lower external magnetic field
- Smaller for a given power rating
- Higher cost in most cases, as winding requires more complex & slower equipment
- Less robust
- Central fixing is either
  - Bolt, large metal washers & rubber pads
  - Bolt & potting resin
- Overtightening the central fixing bolt may short the windings
- Greater inrush current at switch-on.

Autotransformer

An autotransformer has only a single winding, which is tapped at some point along the winding. AC or pulsed voltage is applied across a portion of the winding, and a higher (or lower) voltage is produced across another portion of the same winding. The higher voltage will be connected to the ends of the winding, and the lower voltage from one end to a tap. For example, a transformer with a tap at the centre of the winding can be used with 230 volts across the entire winding, and 115 volts between one end and the tap.

It can be connected to a 230-volt supply to drive 115-volt equipment, or reversed to drive 230-volt equipment from 115 volts. Since the current in the windings is lower, the transformer is smaller, lighter cheaper and more efficient. For voltage ratios not exceeding about 3:1, an autotransformer is cheaper, lighter, smaller and more efficient than an isolating (two-winding) transformer of the same rating. Large three-phase autotransformers are used in electric power distribution systems, for example, to interconnect 33 kV and 66 kV sub-transmission networks.

In practice, transformer losses mean that autotransformers are not perfectly reversible; one designed for stepping down a voltage will deliver slightly less voltage than required if used to step up. The difference is usually slight enough to allow reversal where the actual voltage level is not critical. This is true of isolated winding transformers too.

Variac

By exposing part of the winding coils of an autotransformer, and making the secondary connection through a sliding carbon brush, an autotransformer with a near-continuously variable turns ratio can be obtained, allowing for wide voltage adjustment in very small increments.

Stray Field Transformer

A Stray field transformer has a significant stray field or a (sometimes adjustable) magnetic bypass in its core. It can act as a transformer with inherent current limitation due to its lower tight coupling between the primary and the secondary winding, which is unwanted in much other cases. The output and input currents are low enough to prevent thermal overload under each load condition - even if the secondary is shortened.

Stray field transformers are used for arc welding and high voltage discharge lamps (cold cathode fluorescent lamps, series connected up to 7,5 kV AC working voltage). It acts both as voltage transformer and magnetic ballast.

Polyphase Transformers

For three-phase power, three separate single-phase transformers can be used, or all three phases can be connected to a single polyphase transformer. The three primary windings are connected together and the three secondary windings are connected together. The most common connections are Y-Delta, Delta-Y, Delta-Delta and Y-Y. A vector group indicates the configuration of the windings and the phase angle difference between them.

If a winding is connected to earth (grounded), the earth connection point is usually the centre point of a Y winding. If the secondary is a Delta winding, the ground may be connected to a centre tap on one winding (high leg delta) or one phase may be grounded (corner grounded delta). A special purpose polyphase transformer is the zigzag transformer. There are many possible configurations that may involve more or fewer than six windings and various tap connections.

Resonant Transformers

A resonant transformer operates at the resonant frequency of one or more of its coils and (usually) an external capacitor. The resonant coil, usually the secondary, acts as an inductor, and is connected in series with a capacitor. When the primary coil is driven by a periodic source of alternating current, such as a square or sawtooth wave at the resonant frequency, each pulse of current helps to build up an oscillation in the secondary coil.

Due to resonance, a very high voltage can develop across the secondary, until it is limited by some process such as electrical breakdown. These devices are used to generate high alternating voltages, and the current available can be much larger than that from electrostatic machines such as the Van de Graaff generator or Wimshurst machine.

Examples:

- Tesla coil
- Oudin coil (or Oudin resonator; named after its inventor Paul Oudin)
- D'Arsonval apparatus
- Ignition coil or induction coil used in the ignition system of a petrol engine
- Flyback transformer of a CRT television set or video monitor.
- Electrical breakdown and insulation testing of high voltage equipment and cables. In the latter case, the transformer's secondary is resonated with the cable's capacitance.

Other applications of resonant transformers are as coupling between stages of a superheterodyne receiver, where the selectivity of the receiver is provided by the tuned transformers of the intermediate-frequency amplifiers.

Constant Voltage Transformer

By arranging particular magnetic properties of a transformer core, and installing a ferro-resonant tank circuit (a capacitor and an additional winding), a transformer can be arranged to automatically keep the secondary winding voltage relatively constant for varying primary supply without additional circuitry or manual adjustment.

CVA transformers run hotter than standard power transformers, because regulating action depends on core saturation, which reduces efficiency somewhat. The output waveform is heavily distorted unless careful measures are taken to prevent this. Saturating transformers provide a simple rugged method to stabilize an AC power supply.

Ferrite Core

Ferrite core power transformers are widely used in switched mode power supplies (SMPSUs). The powder core enables high frequency operation, and hence much smaller size to power ratio than laminated iron transformers. Ferrite transformers are not usable as power transformers at mains frequency.

Planar Transformer

Manufacturers etch spiral patterns on a printed circuit board to form the "windings" of a planar transformer. (Manufacturers literally wind pieces of wire on some core or bobbin to form the windings of other kinds of transformers).

Some planar transformers are commercially sold as discrete components — the transformer is the only thing on that printed circuit board.

Other planar transformers are one of many components on one large printed circuit board.

- Much thinner than other transformers, for low-profile applications (even when several PCBs are stacked)
- Almost all use a ferrite planar core

Oil Cooled Transformer

For large transformers used in power distribution or electrical substations, the core and coils of the transformer are immersed in oil which cools and insulates. Oil circulates through ducts in the coil and around the coil and core assembly, moved by convection. The oil is cooled by the outside of the tank in small ratings, and in larger ratings an air-cooled radiator is used.

Where a higher rating is required, or where the transformer is used in a building or underground, oil pumps are used to circulate the oil and an oil-to-water heat exchanger may also be used. Formerly, indoor transformers required to be fire-resistant used PCB liquids; since these are now banned, substitute fire-resistant liquids such as silicone oils are instead used.

Isolating Transformer

Most transformers isolate, meaning the secondary winding is not connected to the primary. But this isn't true of all transformers. However the term 'isolating transformer' is normally applied to mains transformers providing isolation rather than voltage transformation. They are simply 1:1 laminated core transformers. Extra voltage tappings are sometimes included, but to earn the name 'isolating transformer' it is expected that they will usually be used at 1:1 ratio.

Instrument Transformers

Current Transformers

A current transformer (CT) is a measurement device designed to provide a current in its secondary coil proportional to the current flowing in its primary. Current transformers are commonly used in metering and protective relaying in the electrical power industry where they facilitate the safe measurement of large currents, often in the presence of high voltages. The current transformer safely isolates measurement and control circuitry from the high voltages typically present on the circuit being measured.

Current transformers are often constructed by passing a single primary turn (either an insulated cable or an uninsulated bus bar) through a well-insulated toroidal core wrapped with many turns of wire. The CT is typically described by its current ratio from primary to secondary. For example, a 4000:5 CT would provide an output current of 5 amperes when the primary was passing 4000 amperes.

The secondary winding can be single ratio or have several tap points to provide a range of ratios. Care must be taken that the secondary winding is not disconnected from its load while current flows in the primary, as this will produce a dangerously high voltage across the open secondary and may permanently affect the accuracy of the transformer.

Specially constructed wideband CTs are also used, usually with an oscilloscope, to measure high frequency waveforms or pulsed currents within pulsed power systems. One type provides a voltage output that is proportional to the measured current; another, called a Rogowski coil, requires an external integrator in order to provide a proportional output.

Voltage Transformers

Voltage transformers (VTs) or potential transformers (PTs) are another type of instrument transformer, used for metering and protection in high-voltage circuits. They are designed to present negligible load to the supply being measured and to have a precise voltage ratio to accurately step down high voltages so that metering and protective relay equipment can be operated at a lower potential.

Typically the secondary of a voltage transformer is rated for 69 or 120 Volts at rated primary voltage, to match the input ratings of protection relays. The transformer winding high-voltage connection points are typically labelled as H1, H2 (sometimes H0 if it is internally grounded) and X1, X2, and sometimes an X3 tap may be present. Sometimes a second isolated winding (Y1, Y2, Y3) may also be available on the same voltage transformer. The high side (primary) may be connected phase to ground or phase to phase. The low side (secondary) is usually phase to ground. The terminal identifications (H1, X1, Y1, etc.) are often referred to as polarity. This applies to current transformers as well. At any instant terminals with the same suffix numeral have the same polarity and phase. Correct identification of terminals and wiring is essential for proper operation of metering and protection relays.

While VTs were formerly used for all voltages greater than 240V primary, modern meters eliminate the need VTs for most secondary service voltages. VTs are typically used in circuits where the system voltage level is above 600 V. Modern meters eliminate the need of VT's since the voltage remains constant and it is measured in the incoming supply.

Pulse Transformers

A pulse transformer is a transformer that is optimised for transmitting rectangular electrical pulses (that is, pulses with fast rise and fall times and a relatively constant amplitude). Small versions called *signal* types are used in digital logic and telecommunications circuits, often for matching logic drivers to transmission lines.

Medium-sized *power* versions are used in power-control circuits such as camera flash controllers. Larger *power* versions are used in the electrical power distribution industry to interface low-voltage control circuitry to the high-voltage gates of power semiconductors. Special high voltage pulse transformers are also used to generate high power pulses for radar, particle accelerators, or other high energy pulsed power applications.

To minimise distortion of the pulse shape, a pulse transformer needs to have low values of leakage inductance and distributed capacitance, and a high open-circuit inductance. In power-type pulse transformers, a low coupling capacitance (between the primary and secondary) is important to protect the circuitry on the primary side from high-powered transients created by the load.

For the same reason, high insulation resistance and high breakdown voltage are required. A good transient response is necessary to maintain the rectangular pulse shape at the secondary, because a pulse with slow edges would create switching losses in the power semiconductors.

The product of the peak pulse voltage and the duration of the pulse (or more accurately, the voltage-time integral) is often used to characterise pulse transformers. Generally speaking, the larger this product, the larger and more expensive the transformer. Pulse transformers by definition have a duty cycle of less than 1, whatever energy stored in the coil during the pulse must be "dumped" out before the pulse is fired again.

RF Transformers

There are several types of transformer used in radio frequency (RF) work. Steel laminations are not suitable for RF.

Air-Core Transformers

These are used for high frequency work. The lack of a core means very low inductance. Such transformers may be nothing more than a few turns of wire soldered onto a printed circuit board.

Ferrite-Core Transformers

Widely used in intermediate frequency (IF) stages in superheterodyne radio receivers. These are mostly tuned transformers, containing a threaded ferrite slug that is screwed in or out to adjust IF tuning. The transformers are usually canned for stability and to reduce interference.

Transmission-Line Transformers

For radio frequency use, transformers are sometimes made from configurations of transmission line, sometimes bifilar or coaxial cable, wound around ferrite or other types of core. This style of transformer gives an extremely wide bandwidth but only a limited number of ratios (such as 1:9, 1:4 or 1:2) can be achieved with this technique.

The core material increases the inductance dramatically, thereby raising its Q factor. The cores of such transformers help improve performance at the lower frequency end of the band. RF transformers sometimes used a third coil (called a tickler winding) to inject feedback into an earlier (detector) stage in antique regenerative radio receivers.

Baluns

Baluns are transformers designed specifically to connect between balanced and unbalanced circuits. These are sometimes made from configurations of transmission line and sometimes bifilar or coaxial cable and are similar to transmission line transformers in construction and operation.

Audio Transformers

Audio transformers are usually the factor which limit sound quality when used; electronic circuits with wide frequency response and low distortion are relatively simple to design.

Transformers are also used in DI boxes to convert high-impedance instrument signals (e.g. bass guitar) to low impedance signals to enable them to be connected to a microphone input on the mixing console.

A particularly critical component is the output transformer of an audio power amplifier. Valve circuits for quality reproduction have long been produced with no other (inter-stage) audio transformers, but an output transformer is needed to couple the relatively high impedance (up to a few hundred ohms depending upon configuration) of the output valve(s) to the low impedance of a loudspeaker. (The valves can deliver a low current at a high voltage; the speakers require high current at low voltage.) Most solid-state power amplifiers need no output transformer at all.

For good low-frequency response a relatively large iron core is required; high power handling increases the required core size. Good high-frequency response requires carefully designed and implemented windings without excessive leakage inductance or stray capacitance. All this makes for an expensive component. Early transistor audio power amplifiers often had output transformers, but they were eliminated as designers discovered how to design amplifiers without them.

Loudspeaker Transformers

In the same way that transformers are used to create high voltage power transmission circuits that minimize transmission losses, loudspeaker transformers can be used allow many individual loudspeakers to be powered from a single audio circuit operated at higher-than normal loudspeaker voltages. This application is common in industrial public address applications. Such circuits are commonly referred to as constant voltage speaker systems, although the audio waveform is a changing voltage. Such systems are also known by other terms such as 25-, 70- and 100-volt speaker systems, referring to the nominal voltage of the loudspeaker line.

At the audio amplifier, a large audio transformer may be used to step-up the low impedance, low-voltage output of the amplifier to the designed line voltage of the loudspeaker circuit. At the distant loudspeaker location, a smaller step-down transformer returns the voltage and impedance to ordinary loudspeaker levels. The loudspeaker transformers commonly have multiple primary taps, allowing the volume at each speaker to be adjusted in discrete steps.

Output Transformer (valve)

Valve (tube) amplifiers almost always use an output transformer to match the high load impedance requirement of the valves (several kilohms) to a low impedance speaker.

Small Signal Transformers

Moving coil phonograph cartridges produce a very small voltage. In order for this to be amplified with a reasonable signal-noise ratio, a transformer is usually used to convert the voltage to the range of the more common moving-magnet cartridges.

Microphones may also be matched to their load with a small transformer, which is mumetal shielded to minimise noise pickup. These transformers are less widely used today, as transistorized buffers are now cheaper.

'Interstage' and Coupling Transformers

A use for interstage transformers is in the case of push-pull amplifiers where an inverted signal is required. Here two secondary windings wired in opposite polarities may be used to drive the output devices. These phase splitting transformers are not much used today.

Cast Resin Transformers

Cast-resin power transformers have been widely used for a long time.

These transformers have the advantage of easy installation and improved fire behaviour in case of class. This indoor type transformer is totally dry, without cooling oil.

Homemade & Obsolete Transformers

Transformer Kits

Transformers may be wound at home using commercial transformer kits, which contain laminations & bobbin. Or ready made transformers may be disassembled and rewound. These approaches are occasionally used by home constructors, but are usually avoided where possible due to the number of hours required to hand wind a transformer.

100% Homemade

It is possible to make the transformer laminations by hand too. Such transformers are encountered at times in 3rd world countries, using laminations cut from scrap sheet steel, paper slips between the laminations, and string to tie the assembly together. The result works, but is usually noisy due to poor clamping of laminations.

- Picture
- Device in use

Hedgehog

Hedgehog transformers are occasionally encountered in homemade 1920s radios. They are homemade audio interstage coupling transformers. Enamelled copper wire is wound round the central half of the length of a bundle of insulated iron wire (eg florists' wire), to make the windings.

The ends of the iron wires are then bent around the electrical winding to complete the magnetic circuit, and the whole is wrapped with tape or string to hold it together. These were sometimes used when the cost of a ready made transformer could not be justified. Inductance tends to be on the low side, with consequent loss of bass. With the speakers of the day this was no bad thing.

Variocouplers

Variocouplers are rf transformers with 2 windings and variable coupling between the windings. They were standard equipment in 1920s radio sets. Pancake coil variocouplers were common in 1920s radios for variable rf coupling. The 2 planar coils were arranged to swing away from each other and for the angle between them to increase to 90 degrees, thus giving wide variation in coupling. No core was used.

These were mostly used to control reaction. The pancake structure was a means to minimize stray capacitance. In another design of variocoupler, 2 coils were wound on a 2 circular bands, and housed one inside the other, with provision for rotating the inner coil. Coupling varies as one coil is rotated between 0 and 90 degrees from the other. These had higher stray capacitance than the pancake type.

Not Transformers

Finally there are some items often mistaken for transformers, but which are not always transformers. Wall warts: small power supplies with integral mains plug. These can contain a transformer and other circuitry. Most use a laminated iron transformer, but an increasing number now contain a small SMPSU.

These are smaller and much lighter. Halogen lighting transformers: Toroidal transformers are sometimes used for this task, but most halogen 'transformers' are SMPSUs.

Chapter 3

Motors

Electric motors range in size from ones much smaller than the toy motor mounted on your circuit board to motors that pull railroad trains. There is a complex and highly developed technology associated with the large motors used in industry, but most of it is not relevant to the small motors used in behaviour research. Useful laboratory motors usually generate less than I horsepower, and are called fractional horsepower motors. Their use does not require a very profound understanding of the basic principles of motor action, and in this chapter, such principles will only be touched upon.

Behavioural research workers are primarily concerned with the motors' speed characteristics, and these will be the main topic of this chapter. There are only a few considerations which need to be taken into account when choosing an electric motor. The most important ones are the force it will exert, its speed of rotation, and the constancy or adjustability of its speed.

TORQUE

The force that a motor will exert is usually stated in terms of torque, a word that may need some explanation. If a device is to lift an object weighing I pound, it must exert more than I pound of force. It is simply a weight, weighing a little more than 1 pound, on a string that runs over a pulley. The device will also lift a 1-pound object. However, because of the well-known laws of the lever, it is not really correct to give the machine a rating of 1 pound of force.

It exerts 1 pound of force at the place labeled a in the figure, but exerts 3 pounds at b, and ½ pound at c. It would be more completely specified if it were given a rating such as "3 pounds at one third inch." Torque is the term that applies to this sort of combination of force and distance. The torque of the device is 1 pound-inch. This rating means that the device will exert I pound of force at I inch, ½ pound at 2 inches, etc.

In general, the product of the force and the distance equals 1 pound-inch. The mode of movement of this lever is one of rotation around a pivotal point. Any device, such as an electric motor, that has a rotational mode of movement can be most concisely rated in terms of its torque.

For example, the motor needed to lift the weight must be one rated at atleast 10 ounce-inches.

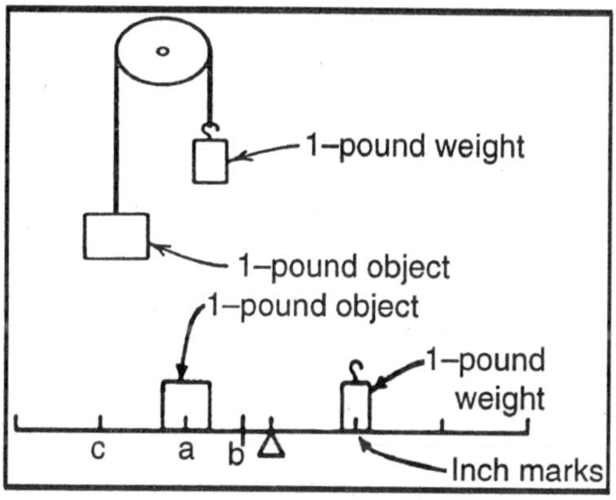

Fig. (*a*) A simple machine to Lift a 1-pound Object. (*b*) Another Machine to Lift a 1-pound Object. The Bar, Marked off in Inches, Pivots About its Centre. The Device Exerts a Torque of I Pound-inch.

A motor rating that is related to torque is horsepower. Horsepower is directly proportional to the product of torque and speed. (A 1horsepower motor will lift a 550-pound weight 1 foot per second.)

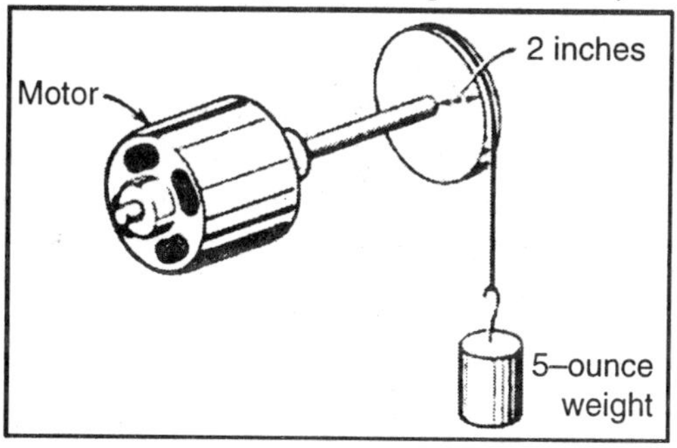

Fig. A Motor and Pulley to Lift a Weight. In Order that the Weight be Lifted, the Motor Must Exert a Torque of More than 10 Ounce-inches.

Most laboratory motors deliver ⅕ horsepower or less, and 1-horsepower electric motors are fairly large. (Vacuum cleaner motors range from about ¾ to 1½ horsepower.)

SPEED CHARACTERISTICS

Electric motors can be classified into three types, according to their speed characteristics. For the purposes of this discussion, the types will be called (1) moderately constant, not very adjustable speed, (2) adjustable speed, and (3) absolutely constant speed.

MODERATELY ADJUSTABLE

Most of the motors used in laboratory equipment run on alternating current at relatively fixed speeds. A typical speed is 1700 revolutions per minute. By using built-in gear boxes, the output shafts of many such motors are made to turn at slower rates. As long as this type of motor is not severely overloaded, it will run at a speed reasonably close to its rated rpm. (The rated speed is usually stamped on the nameplate.)

However, changes in the torque needed to turn the load, and changes in the line voltage, will result in changes in the speed. For this reason such motors should not be used when speed control is critical. Neither should they be used when it is necessary to adjust the speed of rotation. Reducing the input voltage will slow this kind of motor down, but it will also greatly reduce its torque. As a consequence, if the voltage is cut in half, the motor will either run at a very erratic speed or stop altogether. If the label on a motor does not indicate that it belongs to either of the two categories to be discussed below, it probably belongs to this first category.

ADJUSTABLE-SPEED MOTORS

Frequently an experiment requires a motor the speed of which is adjustable and only moderately constant once its level has been set. There is one type of motor that fulfills this requirement very well. It is variously called a series motor, a universal motor, or a sewing machine motor. If any of those terms is stamped on the nameplate of an unknown motor, or if the nameplate simply says a-c/d-c, it is a series motor. The speed of a series motor may be easily adjusted over a ten-to-one range by running the motor from the output of an adjustable voltage transformer. When controlled in this way, the motor will deliver a substantial amount of torque at all speeds.

When a series motor is connected across a source of power, the amount of current it draws is inversely related to its speed of rotation. For example, if a large load is suddenly added to the shaft, slowing it down, the motor will draw a correspondingly large current. If the source of power contains an appreciable internal resistance, this increase in current will cause an increased voltage drop across the resistance, and less voltage will be available to run the motor. The torque will consequently be reduced. Since the internal resistance of an adjustable voltage transformer is quite small, it offers a good means of controlling the speed of a series motor.

However, if a series resistance or a potentiometer is used to control the voltage to the motor, the motor will lose torque and its speed will become unstable when it is reduced. Series motors typically turn at relatively high speeds (5000 to 10,000 rpm) when run at rated voltage, but they may be purchased with built in gear boxes which reduce the maximum rated output speeds to almost any lower value.

CONSTANT-SPEED MOTORS

There are many applications in which the speed of a motor must be absolutely constant, e.g., when a motor is used to time a sequence of stimuli. In these cases, the motor needed is called a synchronous motor. A synchronous motor will run at exactly the speed for which it is rated over a very wide range of applied loads and voltages. Almost all electric clocks, for example, are driven by synchronous motors. In such a motor, the armature is driven by what amounts to a rotating magnetic field, and the rate of rotation of the field depends only upon the physical structure of the motor and upon the *frequency* of the voltage that is driving it.

So long as the line voltage alternates at 60 cycles per second, a synchronous motor will run at a constant speed (except with very great loads or very low voltages, when it wili simply stop). Whereas the line voltage may vary from time to time, the line frequency is controlled with extreme accuracy at the power station. Synchronous motors almost always run at some even submultiple of 3600 rpm (60 revolutions per second). Probably the most common speed for the armature is 1800 rpm, but other armature speeds are available. Gear boxes built into synchronous motors produce output speeds all the way down to I revolution per year.

At any given horsepower or torque, synchronous motors are some what bigger and more expensive than the other types.

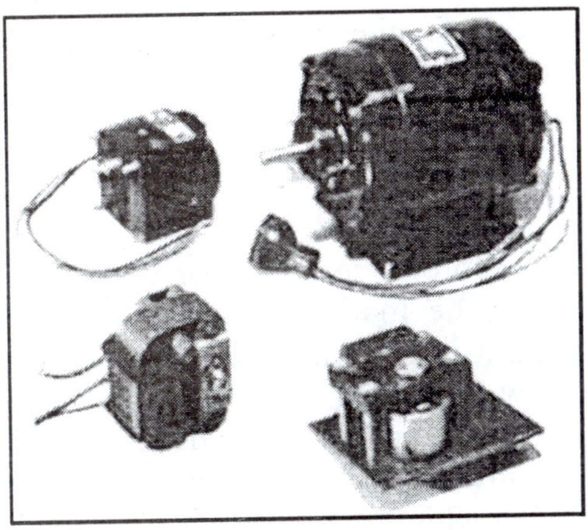

Fig. Various Synchronous Motors.

In the laboratory, they usually fall into one of two groups; large, high-torque motors, for doing real work (e.g., pulling paper through a kymograph), and small, low-torque motors, for timing purposes (e.g., turning a cam that operates a switch to shock a brain every 30 seconds).

REVERSIBLE MOTORS

Most electric motors discussed so far will turn in only one direction, but there are special types whose direction of rotation can be readily reversed. Reversible motors of the first type (moderately constant speed, not very adjustable) are simply called reversible motors. Typically, four wires of different colours emerge from the casing. When two of them are connected together and the other two are connected across the power supply, the motor will run in one direction and, when a different combination is connected, the motor will reverse. The particular combinations are ideosyncratic to individual motors, and such motors always come with instructions on how to connect them up.

Reversible synchronous motors are also available but they are less common. They usually consist of two different motors connected to one output shaft, one a clockwise and the other a counterclockwise driver. The direction of rotation of the shaft depends upon which driver has voltage applied to it. The direction of rotation of standard series motors cannot be reversed. When a motor is needed whose speed is adjustable and whose direction of rotation is reversible, still a different kind of motor is available.

The field of this kind of motor is supplied by a set of permanent magnets instead of the electromagnetic field present in the motors discussed previously. It is called a permanent-magnet motor. The one on your circuit board is of this type. Reversing the direction of current flow through a permanent magnet motor reverses the direction of rotation, and lowering the voltage reduces the speed. Permanent-magnet motors run on direct current only, and, except for the very small ones that will not do much work, they draw too much current to be powered by dry-cell batteries. Therefore, they are inconvenient to use unless a source of direct current is available in your laboratory.

GOVERNOR-CONTROLLED MOTORS

Ordinarily, the speed of a series motor varies with small changes in the line voltage or the load. However, some series motors are manufactured with built-in governors to maintain a constant speed. It will be sufficient now to say that three types of governor-controlled series motors are available, one with a fixed speed, another whose speed can be varied by an adjustment of the governor when the motor is turned off, and a third type in which the governor can be adjusted while the motor is actually running.

SPEED CHARACTERISTICS

The following is a list of the kinds of motors which fit specific requirements:

- For very constant speed on a-c power:
  - Synchronous.
  - Governor-controlled series.
- For constant speed on direct current:
  - Governor-controlled series.
- For adjustable speed on alternating current:
  - Series (universal, sewing-machine).
  - Series with adjustable governor (for constant speed at each setting).
- For adjustable speed on direct current:
  - Series.
  - Series with adjustable governor.
  - Permanent magnet.
- For reversible direction of rotation on alternating current:
  - Reversible alternating current.
  - Reversible synchronous (for very constant speed).
- For reversible direction on direct current.
  - Permanent magnet.

MISCELLANEOUS INFORMATION

There are some additional details concerning these motors which, though of less general relevance, are still worth noting.

High Starting Torque

Series motors have one outstanding characteristic that has not been mentioned. For a fixed input voltage, their output torque is inversely related to their speed. In other words, the series motor exerts its greatest torque when it is first turned on and is just starting up. This is not the case for the other motors that have been discussed. Should you need a motor that starts up quickly under a heavy load and does not need a push to get it going, use a series motor.

Sparking

When an electric motor runs, electric currents circulate in its armature. In most a-c motors, there is no direct electric connection between the armature and the power supply; current is electromagnetically induced into the armature.

(For this reason, such motors are called induction motors.) But in d-c and a-c/d-c (series) motors, the armature current is brought in through a set of moving contacts called the brushes and commutator.

As the armature turns, these contacts are rapidly made and broken, and sparks tend to jump at each break. You can see these sparks if you look at the spinning shaft of your toy motor in a darkened room. The sparks act as small radio broadcasting stations, and play havoc with any sensitive electronic apparatus that may be operating nearby. For this reason, d-c and a-c/d-c motors should not be used near bioelectric recording equipment.

Chapter 4

Alternating Currents

An alternating current (AC) is an electric current whose direction reverses cyclically, as opposed to direct current, whose direction remains constant. The usual waveform of an AC power circuit is a sine wave, as this results in the most efficient transmission of energy. However in certain applications different waveforms are used, such as triangular or square waves.

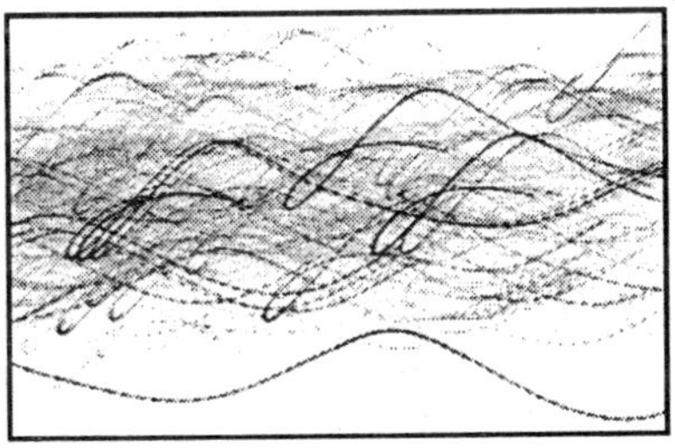

Fig. City lights viewed in a motion blurred exposure. The AC blinking causes the lines to be dotted rather than continuous.

Used generically, *AC* refers to the form in which electricity is delivered to businesses and residences. However, audio and radio signals carried on electrical wires are also examples of alternating current. In these applications, an important goal is often the recovery of information encoded (or modulated) onto the AC signal.

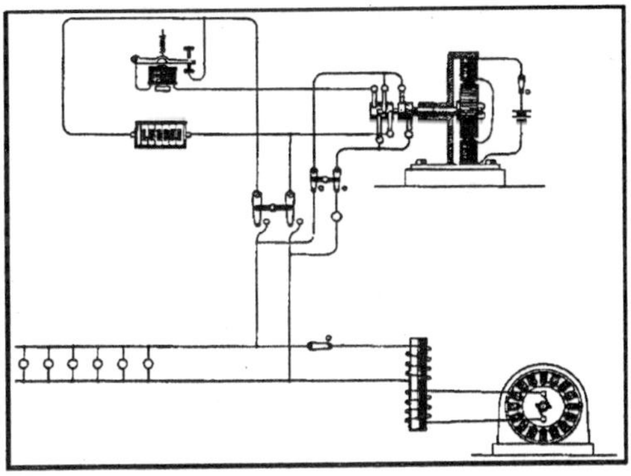

Fig. Westinghouse Early AC System 1887.

William Stanley, Jr. designed one of the first practical devices to transfer AC power efficiently between isolated circuits. Using pairs of coils wound on a common iron core, his design, called an induction coil, was an early transformer. The AC power system used today developed rapidly after 1886, and includes key concepts by Nikola Tesla, who subsequently sold his patent to George Westinghouse. Lucien Gaulard, John Dixon Gibbs, Carl Wilhelm Siemens and others contributed subsequently to this field.

AC systems overcame the limitations of the direct current system used by Thomas Edison to distribute electricity efficiently over long distances even though Edison attempted to discredit alternating current as too dangerous during the War of Currents. The first commercial power plant using three-phase alternating current was at the Mill Creek hydroelectric plant near Redlands, California in 1893 designed by Almirian Decker. Decker's design incorporated 10,000 volt three-phase transmission and established the standards for the complete system of generation, transmission and motors used today.

Alternating current circuit theory evolved rapidly in the latter part of the 19th and early 20th century. Notable contributors to the theoretical basis of alternating current calculations include Charles Steinmetz, James Clerk Maxwell, Oliver Heaviside, and many others. Calculations in unbalanced three-phase systems were simplified by the symmetrical components methods discussed by Charles Legeyt Fortescue in 1918.

TRANSMISSION, DISTRIBUTION, AND DOMESTIC POWER SUPPLY

AC voltage may be increased or decreased with a transformer. Use of a higher voltage leads to significantly more efficient transmission of power. The power losses in a conductor are a product of the square of the current and the resistance of the conductor, described by the formula. This means that when transmitting a fixed power on a given wire, if the current is doubled, the power loss will be four times greater.

Since the power transmitted is equal to the product of the current, the voltage and the cosine of the phase difference φ ($P = IV\cos\varphi$), the same amount of power can be transmitted with a lower current by increasing the voltage. Therefore it is advantageous when transmitting large amounts of power to distribute the power with high voltages (often hundreds of kilovolts).

However, high voltages also have disadvantages, the main one being the increased insulation required, and generally increased difficulty in their safe handling. In a power plant, power is generated at a convenient voltage for the design of a generator, and then stepped up to a high voltage for transmission. Near the loads, the transmission voltage is stepped down to the voltages used by equipment. Consumer voltages vary depending on the country and size of load, but generally motors and lighting are built to use up to a few hundred volts between phases.

The utilization voltage delivered to equipment such as lighting and motor loads is standardized, with an allowable range of voltage over which equipment is expected to operate. Standard power utilization voltages and percentage tolerance vary in the different mains power systems found in the world. Modern high-voltage, direct-current electric power transmission systems contrast with the more common alternating-current systems as a means for the bulk transmission of electrical power over long distances. HVDC systems tend to be more expensive and less efficient than transformers.

Transmission with high voltage direct current was not feasible when Edison, Westinghouse and Tesla were designing their power systems, since there was then no way to economically convert AC power to DC and back again at the necessary voltages. Three-phase electrical generation is very common. Three separate coils in the generator stator are physically offset by an angle of 120° to each other. Three current waveforms are produced that are equal in magnitude and 120° out of phase to each other.

If the load on a three-phase system is balanced equally among the phases, no current flows through the neutral point. Even in the worst-case unbalanced (linear) load, the neutral current will not exceed the highest of the phase currents. Non-linear loads (e.g. computers) may require an oversized neutral bus and neutral conductor in the upstream distribution panel to handle harmonics. Harmonics can cause neutral conductor current levels to exceed that of one or all phase conductors.

For three-phase at utilization voltages a four-wire system is often used. When stepping down three-phase, a transformer with a Delta primary and a Star secondary is often used so there is no need for a neutral on the supply side. For smaller customers (just how small varies by country and age of the installation) only a single phase and the neutral or two phases and the neutral are taken to the property. For larger installations all three phases and the neutral are taken to the main distribution panel. From the three-phase main panel, both single and three-phase circuits may lead off.

Three-wire single phase systems, with a single centre-tapped transformer giving two live conductors, is a common distribution scheme for residential and small commercial buildings in North America. This arrangement is sometimes incorrectly referred to as "two phase". A similar method is used for a different reason on construction sites in the UK. Small power tools and lighting are supposed to be supplied by a local centre-tapped transformer with a voltage of 55V between each power conductor and the earth. This significantly reduces the risk of electric shock in the event that one of the live conductors becomes exposed through an equipment fault whilst still allowing a reasonable voltage for running the tools.

A third wire, called the bond wire, is often connected between non-current-carrying metal enclosures and earth ground. This conductor provides protection from electric shock due to accidental contact of circuit conductors with the metal chassis of portable appliances and tools. Bonding all non-current-carrying metal parts into one complete system ensures there is always a low impedance path to ground sufficient to carry any fault current for as long as it takes for the system to clear the fault. This low impedance path allows the maximum amount of fault current, causing the overcurrent protection device (Breakers, fuses) to trip or burn out as quickly as possible, returning the electrical system to a safe state. All bond wires are bonded to ground at the main service panel, as is the Neutral/Identified conductor if present.

ELECTRICITY DISTRIBUTION

Electricity distribution is the penultimate stage in the delivery (before retail) of electricity to end users. It is generally considered to include medium-voltage (less than 50 kV) power lines, electrical substations and pole-mounted transformers, low-voltage (less than 1000 V) distribution wiring and sometimes electricity meters.

In the early days of electricity distribution, direct current DC generators were connected to loads at the same voltage. The generation, transmission and loads had to be of the same voltage because there was no way of changing DC voltage levels, other than inefficient motor-generator sets. Low DC voltages were used (on the order of 100 volts) since that was a practical voltage for incandescent lamps, which were then the primary electrical load. The low voltage also required less insulation to be safely distributed within buildings.

The losses in a cable are proportional to the square of the current, the length of the cable, and the resistivity of the material, and are inversely proportional to cross-sectional area. Early transmission networks were already using copper, which is one of the best economically feasible conductors for this application. To reduce the current and copper required for a given quantity of power transmitted would require a higher transmission voltage, but no convenient efficient method existed to change the voltage level of DC power circuits. To keep losses to an economically practical level the Edison DC system needed thick cables and local generators. Early DC generating plants needed to be within about 1.5 miles (2.4 km) of the farthest customer to avoid the need for excessively large and expensive conductors.

The adoption of alternating current (AC) for electricity generation following the War of Currents dramatically changed the situation. Power transformers, installed at power stations, could be used to raise the voltage from the generators, and transformers at local substations reduced it to supply loads. Increasing the voltage reduced the current in the transmission and distribution lines and hence the size of conductors required and distribution losses incurred. This made it more economical to distribute power over long distances. Generators (such as hydroelectric sites) could be located far from the loads.

In North America, early distribution systems used a voltage of 2200 volts corner-grounded delta. Over time, this was gradually increased to 2400 volts. As cities grew, most 2400 volt systems were upgraded to 2400/4160 volt, three-phase systems. In three phase networks that permit connections between phase and neutral, both the phase-to-phase voltage (4160, in this example) and the phase-to-neutral voltage are given; if only one value is shown, the network does not serve single-phase loads connected phase-to-neutral. Some city and suburban distribution systems continue to use this range of voltages, but most have been converted to 7200/12470Y, 7620/13200Y, 14400/24940Y, and 19920/34500Y.

European systems used 3300 volts to ground, in support of the 220/380Y volt power systems used in those countries. In the UK, urban systems progressed to 6.6 kV and then 11 kV (phase to phase), the most common distribution voltage. North American and European power distribution systems also differ in that North American systems tend to have a greater number of low-voltage, step-down transformers located close to customers' premises. For example, in the US a pole-mounted transformer in a suburban setting may supply 1-3 houses, whereas in the UK a typical urban or suburban low-voltage substation would normally be rated between 315kVA and 1000kVA (1MVA) and supply a whole neighbourhood.

This is because the higher voltage used in Europe (415V vs 230V) may be carried over a greater distance with acceptable power loss. An advantage of the North American setup is that failure or maintenance on a single transformer will only affect a few customers. Advantages of the UK setup are that the transformers may be fewer, larger and more efficient, and due to diversity there need be less spare capacity in the transformers, reducing power wastage. In North American city areas with many customers per unit area, network distribution will be used, with multiple transformers and low-voltage busses interconnected over several city blocks.

Rural Electrification systems, in contrast to urban systems, tend to use higher voltages because of the longer distances covered by those distribution lines. 7200, 12470, 25000, and 34500 volt distribution is common in the United States; 11 kV and 33 kV are common in the UK, New Zealand and Australia; 11 kV and 22 kV are common in South Africa. Other voltages are occasionally used. In New Zealand, Australia, Saskatchewan, Canada, and South Africa, single wire earth return systems (SWER) are used to electrify remote rural areas.

While power electronics now allow for conversion between DC voltage levels, AC is still used in distribution due to the economy, efficiency and reliability of transformers. High-voltage DC is used for transmission of large blocks of power over long distances, or for interconnecting adjacent AC networks, but not for distribution to customers.

Distribution Network Configurations

Distribution networks are typically of two types, radial or interconnected. A radial network leaves the station and passes through the network area with no normal connection to any other supply. This is typical of long rural lines with isolated load areas. An interconnected network is generally found in more urban areas and will have multiple connections to other points of supply.

These points of connection are normally open but allow various configurations by the operating utility by closing and opening switches. Operation of these switches may be by remote control from a control centre or by a lineman. The benefit of the interconnected model is that in the event of a fault or required maintenance a small area of network can be isolated and the remainder kept on supply.

Within these networks there may be a mix of overhead line construction utilizing traditional utility poles and wires and, increasingly, underground construction with cables and indoor or cabinet substations. However, underground distribution is significantly more expensive than overhead construction. In part to reduce this cost, underground power lines are sometimes co-located with other utility lines in what are called Common utility ducts. Distribution feeders emanating from a substation are generally controlled by a circuit breaker which will open when a fault is detected. Automatic Circuit Reclosers may be installed to further segregate the feeder thus minimizing the impact of faults.

Long feeders experience voltage drop requiring capacitors or voltage regulators to be installed. Characteristics of the supply given to customers are generally mandated by contract between the supplier and customer. Variables of the supply include:

- AC or DC - Virtually all public electricity supplies are AC today. Users of large amounts of DC power such as some electric railways, telephone exchanges and industrial processes such as aluminium smelting usually either operate their own or have adjacent dedicated generating equipment, or use rectifiers to derive DC from the public AC supply

- Voltage, including tolerance (usually +10 or -15 percentage)
- Frequency, commonly 50 & 60 Hz, 16-2/3 Hz for some railways and, in a few older industrial and mining locations, 25 Hz.
- Phase configuration (single phase, polyphase including two phase and three phase)
- Maximum demand (usually measured as the largest amount of power delivered within a 15 or 30 minute period during a billing period)
- Load Factor, expressed as a ratio of average load to peak load over a period of time. Load factor indicates the degree of effective utilization of equipment (and capital investment) of distribution line or system.
- Power factor of connected load
- Earthing arrangements - TT, TN-S, TN-C-S or TN-C
- Maximum prospective short circuit current
- Maximum level and frequency of occurrence of transients

Modern Distribution Systems

The modern distribution system begins as the primary circuit leaves the sub-station and ends as the secondary service enters the customer's meter socket. A variety of methods, materials, and equipment are used among the various utility companies, but the end result is similar. First, the energy leaves the sub-station in a primary circuit, usually with all three phases.

The most common type of primary is known as a wye configuration (so named because of the shape of a "Y".) The wye configuration includes 3 phases (represented by the three outer parts of the "Y") and a neutral (represented by the centre of the "Y".) The neutral is grounded both at the substation and at every power pole. In a typical 12470Y/7200 volt system, the pole mount transformer's primary winding is rated for 7200 volts and is connected across one phase of power and the neutral.

The primary and secondary (low voltage) neutrals are bonded (connected) together to provide a path to blow the primary fuse if any fault occurs that allows primary voltage to enter the secondary lines. An example of this type of fault would be a primary phase falling across the secondary lines. Another example would be some type of fault in the transformer itself. The other type of primary configuration is known as delta. This method is older and less common. Delta is so named because of the shape of the Greek letter delta, a triangle. Delta has only 3 phases and no neutral. In delta there is only a single voltage, between two phases (phase to phase), while in wye there are two voltages, between two phases and between a phase and neutral (phase to neutral). Wye primary is safer because if one phase becomes grounded, that is, makes connection to the ground through a person, tree, or other object, it should trip out the fused cutout similar to a household circuit breaker tripping.

In delta, if a phase makes connection to ground it will continue to function normally. It takes two or three phases to make connection to ground before the fused cutouts will open the circuit. The voltage for this configuration is usually 4800 volts. Transformers are sometimes used to step down from 7200 or 7600 volts to 4800 volts or to step up from 4800 volts to 7200 or 7600 volts. When the voltage is stepped up, a neutral is created by bonding one leg of the 7200/7600 side to ground.

This is commonly used to power single phase underground services or whole housing developments that are built in 4800 volt delta distribution areas. Step downs are used in areas that have been upgraded to a 7200/12500Y or 7600/13200Y and the power company chooses to leave a section as a 4800 volt setup. Sometimes power companies choose to leave sections of a distribution grid as 4800 volts because this setup is less likely to trip fuses or reclosers in heavily wooded areas where trees come into contact with lines.

Economic and Political

Traditionally the electricity industry has been a publicly owned institution but starting in the 1970s nations began the process of deregulation and privatisation, leading to electricity markets. A major focus of these was the elimination of the former so called *natural monopoly* of generation, transmission, and distribution. As a consequence, electricity has become more of a commodity. The separation has also led to the development of new terminology to describe the business units, *e.g.* line company, wires business and network company.

AC POWER SUPPLY FREQUENCIES

The frequency of the electrical system varies by country; most electric power is generated at either 50 or 60 Hz. Some countries have a mixture of 50 Hz and 60 Hz supplies, notably Japan. A low frequency eases the design of low speed electric motors, particularly for hoisting, crushing and rolling applications, and commutator-type traction motors for applications such as railways, but also causes a noticeable flicker in incandescent lighting and objectionable flicker of fluorescent lamps. $16\frac{2}{3}$ Hz power is still used in some European rail systems, such as in Austria, Germany, Norway, Sweden and Switzerland.

The use of lower frequencies also provided the advantage of lower impedance losses, which are proportional to frequency. The original Niagara

Falls generators were built to produce 25 Hz power, as a compromise between low frequency for traction and heavy induction motors, while still allowing incandescent lighting to operate (although with noticeable flicker); most of the 25 Hz residential and commercial customers for Niagara Falls power were converted to 60 Hz by the late 1950s, although some 25 Hz industrial customers still existed as of the start of the 21st century. Off-shore, military, textile industry, marine, computer mainframe, aircraft, and spacecraft applications sometimes use 400 Hz, for benefits of reduced weight of apparatus or higher motor speeds.s

Effects at High Frequencies

A direct, constant current flows uniformly throughout the cross-section of the (uniform) wire that carries it. With alternating current of any frequency, the current is forced towards the outer surface of the wire, and away from the centre. This is because an electric charge which accelerates (as is the case of an alternating current) radiates electromagnetic waves, and materials of high conductivity (the metal which makes up the wire) do not allow propagation of electromagnetic waves. This phenomenon is called skin effect.

At very high frequencies the current no longer flows *in* the wire, but effectively flows *on* the surface of the wire, within a thickness of a few skin depths. The skin depth is the thickness at which the current density is reduced by 63%. Even at relatively low frequencies used for high power transmission (50–60 Hz), non-uniform distribution of current still occurs in sufficiently thick conductors. For example, the skin depth of a copper conductor is approximately 8.57 mm at 60 Hz, so high current conductors are usually hollow to reduce their mass and cost.

Since the current tends to flow in the periphery of conductors, the effective cross-section of the conductor is reduced. This increases the effective *AC* resistance of the conductor, since resistance is inversely proportional to the cross-sectional area in which the current actually flows. The AC resistance often is many times higher than the *DC* resistance, causing a much higher energy loss due to ohmic heating (also called IR loss).

Techniques for Reducing AC Resistance

For low to medium frequencies, conductors can be divided into stranded wires, each insulated from one other, and the individual strands specially arranged to change their relative position within the conductor bundle. Wire constructed using this technique is called Litz wire.

This measure helps to partially mitigate skin effect by forcing more equal current flow throughout the total cross section of the stranded conductors. Litz wire is used for making high Q inductors, reducing losses in flexible conductors carrying very high currents at power frequencies, and in the windings of devices carrying higher radio frequency current (up to hundreds of kilohertz), such as switch-mode power supplies and radio frequency transformers.

Techniques for Reducing Radiation Loss

As written above, an alternating current is made of electric charge under periodic acceleration, which causes radiation of electromagnetic waves. Energy that is radiated represents a loss. Depending on the frequency, different techniques are used to minimize the loss due to radiation.

Twisted Pairs

At frequencies up to about 1 GHz, wires are paired together in cabling to form a twisted pair in order to reduce losses due to electromagnetic radiation and inductive coupling. A twisted pair must be used with a balanced signalling system, where the two wires carry equal but opposite currents. The result is that each wire in the twisted pair radiates a signa that is effectively cancelled by the other wire, resulting in almost no electromagnetic radiation.

Coaxial Cables

At frequencies above 1 GHz, unshielded wires of practical dimensions lose too much energy to radiation, so coaxial cables are used instead. A coaxial cable has a conductive wire inside a conductive tube. The current flowing on the inner conductor is equal and opposite to the current flowing on the inner surface of the outer tube. This causes the electromagnetic field to be completely contained within the tube, and (ideally) no energy is radiated or coupled outside the tube.

Coaxial cables have acceptably small losses for frequencies up to about 20 GHz. For microwave frequencies greater than 20 GHz, the dielectric losses (due mainly to the dissipation factor of the dielectric layer which separates the inner wire from the outer tube) become too large, making waveguides a more efficient medium for transmitting energy.

WAVEGUIDES

Waveguides are similar to coax cables, as both consist of tubes, with the biggest difference being that the waveguide has no inner conductor. Waveguides can have any arbitrary cross section, but rectangular cross sections are the most common. Because waveguides do not have an inner conductor to carry a return current, waveguides cannot deliver energy by means of an electric current, but rather by means of a *guided* electromagnetic field.

Although surface currents do flow on the inner walls of the waveguides, those surface currents do not carry power. Power is carried by the guided electromagnetic fields. The surface currents are set up by the guided electromagnetic fields and have the effect of keeping the fields inside the waveguide and preventing leakage of the fields to the space outside the waveguide.

Waveguides have dimensions comparable to the wavelength of the alternating current to be transmitted, so they are only feasible at microwave frequencies. In addition to this mechanical feasibility, electrical resistance of the non-ideal metals forming the walls of the waveguide cause dissipation of power (surface currents flowing on lossy conductors dissipate power). At higher frequencies, the power lost to this dissipation becomes unacceptably large.

Fibre Optics

At frequencies greater than 200 GHz, waveguide dimensions become impractically small, and the ohmic losses in the waveguide walls become large. Instead, fibre optics, which are a form of dielectric waveguides, can be used. For such frequencies, the concepts of voltages and currents are no longer used.

The existence of two types of electric current had been known since the dawn of the electrical age. Hippolyte Pixii's generator of 1832 had had a rotating magnetic field that cut the conductors first in one direction and then in the other. This produced a series of interrupted electrical pulses as the magnets swung back and forth; the flow of electricity reached a maximum in one direction, then decreased, moved toward a maximum in the other direction, and reversed again to begin a new cycle.

Such an "alternating" current did not seem useful to the early experimenters. Ampθre had suggested a simple device called a commutator, which converted the oscillating or alternating current into a non-pulsating "direct" current, and commutators became standard features on the dynamos of the are light era. Edison's first generating machines were direct-current dynamos, but his early commitment to d.c. was accidental, to a large extent. For the purpose of providing incandescent lighting, it did not matter whether Edison used d.c. or a.c.

The dynamo technology of 1878 was almost exclusively a d.c. technology, and so he chose to make no departure from it. Before long, though, he was irrevocably wedded to the d.c. concept as an essential part of his system. What was important to Edison as he created his system was not the type of current he used, but the voltage. The Edison system was based on power that traveled at 110 volts all the way from the generating station to the consumer. The great advantage of this was its safety.

Current at 110 volts and at the low amperage sufficient for electric lighting could do no harm; a customer who came in contact with an unshielded line would receive a mild shock, no more. The danger of fire was low, too. Sensitive to the problems inherent in getting a new kind of illumination across to a suspicious public, Edison wished to avoid any possibility of spectacular electrical disasters, so he opted for the safest form of distribution.

However, low voltages had a built-in drawback: they limited the distance of transmission. Current moving with such a modest force could be carried no more than a mile or two from the generating station; then the drop in voltage became so great that the outlying lights dimmed. In addition, much of the power would be dissipated in the form of heat. Under Edison's low voltage transmission system, Edison's own great concept of the central station was seriously impaired, for 110-volt service required a chain of power stations at close intervals in order to supply any large city with electricity. At best an Edison station could serve some sixteen square miles.

Ideally, the solution was to have power leave the generating station at a high voltage, and then to step it down to a safe 110 volts before it reached the customer. But this was technically impossible on the Edison system. The young Edison would have looked for some way to *make* it possible; the middle-aged Edison simply accepted the situation as a regrettable innate flaw in his design, and left it at that.

But there was a way—with alternating current.

The germ of the long-distance a.c. transmission system lay in the original work of Faraday. His first successful experiment had made use of two coils wound on an iron core; by stopping or starting a current in one coil, he induced a current in the other. Although he did not realize it, the basis of the transformer, or voltage-changer, could be found in this experiment.

Seven years later, in 1838, the American Joseph Henry used Faraday's two-coil arrangement as a step-down transformer. One coil consisted of many turns of wire, the other of only a few; high-voltage power entered the first coil, and was stepped down to a low voltage in the second.

The transformer received little further attention until 1882, when Lucien Gaulard of France and John D. Gibbs of England produced a transformer specifically designed to reduce the voltage of a.c. power en route from the generator to the consumer. Gaulard and Gibbs obtained an English patent on their work in 1883, but in practice it had serious mechanical defects and could not be used without modification.

The modifications were quick in coming. In 1885 three Hungarian inventors found a way to employ the Gaulard-Gibbs transformer in long-distance power installations. Their device, called the "Z.B.D. transformer" after the initials of its inventors, went into immediate use in Europe. It permitted power first to be boosted to very high voltage as it left the generator, then to be stepped down again as it reached the consumer—with remarkably little energy loss over great distances. The higher the voltage used in transmission, the less the loss on the line.

Rights to the Z.B.D. system were offered to Edison in 1886. He sent Upton to Europe to inspect the equipment, and at Upton's suggestion Edison paid $5000 for a three-year option, against a total price of $20,000. But Edison could not bring himself to veer from his low-voltage d.c. transmission system. In November, 1886 he received a report from a Berlin engineering firm to the effect that the Z.B.D. system was troublesome, costly to install, and dangerous to work with. That was all Edison needed to hear. He allowed the option to lapse, and angrily rejected the suggestions of his associates that he give the Hun garian transformer a further trial.

Edison's adherence to his own original system had become something deeply personal, a matter for the emotions, not the reason. He clung to his 110-volt concept despite all advice to the contrary. Yet younger men all over Europe and America were excited by the new development. High-voltage transmission suddenly was the technological darling of the moment. Of course, it would require a changeover from d.c. to a.c. power. Since transformers functioned only with a changing current, the output of d.c. generators could be neither stepped up nor stepped down with transformers. High voltages required the use of a.c. equipment.

And that produced certain new problems. Alternating current could not be used for driving electric motors; it seemed to demand two or three times as much coal per kilowatt hour as direct current; and it rendered obsolete most of the existing capital plant of the power industry. Nevertheless, the forward looking men of that industry were confident that someone would invent a feasible a.c. electric motor and that the generating cost of a.c. could be made comparable to that of d.c. As for the obsolescence of existing equipment, that would cease to matter in time. Alternating current was sure to win universal acceptance.

Edison—nearly alone—fought the new trend.

His option on the Z.B.D. patents gave him the exclusive right to use the new transformer in the United States until 1889. Since that was the case, rival manufacturers had no choice but to go back to the defective Gaulard-Gibbs device and attempt to modify it themselves in a way that circumvented the Hungarians. The most eager of these rivals was George Westinghouse of Pittsburgh, who had gone into the electric business in 1885.

Westinghouse, one year Edison's senior, had founded his industrial fortunes on his invention of the air brake for railways. Bold, ambitious, even reckless, he dreamed of an electrical empire to rival Edison's, and by liberal use of cash had attracted some of the best engineers of the day away from the ever more conservative Edison organization.

The new Westinghouse Electrical & Manufacturing Company sought to steal a march on Edison by getting control of that segment of the electrical field that Edison scorned: high-voltage alternating current equipment of 1,000 volts or more. For that Westinghouse needed a practicable transformer.

Westinghouse bought the American rights to the Gaulard Gibbs transformer and assigned the job of making it work properly to a young engineering consultant, William A. Stanley. The flaw in the Gaulard-Gibbs system was that it employed a number of transformers to be connected in series, so that they could not work independently. Stanley found a patentable way of connecting transformers in parallel and making each transformer automatically regulate itself, thus permitting independent control.

On March 6, 1886, the first commercial alternating current lighting system went into operation in Great Barrington, Massachusetts, using the Stanley transformers. At Great Barrington Stanley generated power at 500 volts just outside town, stepped it up to 3000 volts for transmission, then lowered it to 500 volts again for use in lighting. The transmission distance was only 4000 feet, but the principle was established; Westinghouse saw no reason why the Stanley system could not be used to send power cheaply for distances of many miles.

Westinghouse, in 1886, also purchased the United States Electric Lighting Company, which had control of the contested Maxim incandescent lamp patents. Now he proposed to compete in both spheres of the electrical industry. Through Westinghouse Electric he would manufacture high-voltage equipment; through United States Electric he would set up power companies that, by taking advantage of the economies of long distance transmission, would outstrip the Edison companies. Where he could, Westinghouse would rely on the non-Edison technology of a.c. Where he could not, he intended to invade the Edison patents, whose validity was under attack in the courts. In an advertisement exceptional for its blandly self congratulatory tone Westinghouse declared:

> We regard it as fortunate that we have deferred entering the electrical field until the present moment. Having thus profited by the public experience of others, we enter ourselves for competition, hampered by a minimum of expense for experimental outlay.... In short, our organization is free, in large measure, of the load with which [other] electrical enterprises seem to be encumbered. The fruit of this... we propose to share with the customer.

That is, having allowed Edison to bear the burden of developing the electrical industry, Westinghouse planned to undersell him by making use of his pioneering work. In March of 1887 Westinghouse began to move into New York City, Edison's home grounds, with an a.c. incandescent lighting system. He organized the Safety Electric Light and Power Company and licensed it to use the Westinghouse system of generation and distribution of electricity by alternating current.

Safety Electric began to buy into an existing are light company, United States Illuminating, which held several important street and park contracts. Soon Safety Electric controlled United States Illuminating; it changed its name in 1889 to United Electric Light and Power Company and began to spread the Westinghouse system into those parts of Manhattan that Edison Illuminating had not yet reached.

Naturally the older company fought back. In 1888 it denounced the Westinghouse system in a pamphlet that declared:

- They cannot make it safe.
- They cannot make it reliable.
- They cannot make it run twelve 16-candlepower lamps per horsepower.
- They cannot make its lamps even.
- They cannot make its lamps last a reasonable time.
- They cannot make it sell by meter.
- They cannot make it run motors.

Some of these charges were true at the time; others, particularly the accusation of unsafeness, were not. But the men who had made such a deep financial and emotional investment in d.c. could not abide the thought that their system was obsolescent.

Edison himself, though no longer directly concerned with the fortunes of the Edison Illuminating Company, was seriously perturbed by the spread of a.c. It was an attack on his technical judgment, he felt; he was worried, also, about Westinghouse's talk of using power at 5,000 or even 10,000 volts. Contemporary insulating techniques did not seem adequate for such voltages, Edison felt. To Edward Johnson he wrote:

Just as certain as death Westinghouse will kill a customer within 6 months after he puts in a system of any size. He has got a new thing and it will require a great deal of experimenting to get it working practically. It will never be free from danger....

None of his plans worry me in the least; only thing that disturbs me is that Westinghouse is a great man for flooding the country with agents and travelers. He is ubiquitous and will form numerous companies before we know anything about it.

Edison's earlier rival, Thomson-Houston Electric, was also toying with a.c. Elihu Thomson had constructed a rudimentary transformer in 1879, but he did nothing with it then because, like everyone else at the time, he was building d.c. dynamos. After the success of the three Hungarians provoked interest in a.c., Thomson hastily filed an application in November of 1885 for a patent on his transformer. But several other groups, among them the Gaulard-Gibbs team, contested the application and Thomson's patent was not granted for nearly twenty years, only to be thrown out anyway on its first court test.

Patent or not, Thomson saw the advantages of a long-distance a.c. transmission system, and continued his research. By the spring of 1887 Thomson-Houston was manufacturing a.c. dynamos—"alternators," as they were called—and transformers. However, Thomson was perturbed by the supposed danger of bringing high-voltage lines to the consumers, and so Thomson Houston held back from any really enthusiastic promotion of a.c. for several years, leaving a clear field to Westinghouse while it worked on protective devices.

The problem of the electric motor also hindered the spread of a.c. in 1887. All existing motors worked on d.c., and d.c. alone; no one yet knew a way of adapting them for a.c. use. Since the electrical industry was coming to depend for its revenues more and more on industrial power use, less and less on lighting, a practical a.c. motor was essential if the new system were to last.

Elihu Thomson built one of the first a.c. motors in May of 1887. It had a laminated magnet consisting of many disks of metal pressed together, and a laminated armature with three coils. The Thomson motor attracted considerable attention when it was displayed before a scientific group, but it did not go into production at once, and the aggressive Westinghouse gained a leading position with the a.c. motor of Nikola Tesla.

Tesla, a towering, mystical genius out of Yugoslavia, had once been part of the Edison organization but had slipped away. As an electrical inventor he was extraordinary, very likely superior in vision even to Edison. But he was a romantic, brooding figure of little practical awareness who could never have become a permanent part of the Edison research team. There was room in Edison's entourage for only one genius at a time; besides, Tesla believed passionately in the alternating current.

In 1882, in Budapest, Tesla was seized with a vision of an alternating-current motor: a rotating magnetic field that carried the armature around with it. The Tesla motor employed two alternating currents at once; the currents were equal in the frequency of their cycle of minimum-to-maximum amplitude, but were out of phase relative to each other. With one current or the other always at a maximum, the motor worked continuously. It was a briiliant idea, a landmark in engineering. But in 1882 there was no a.c. power source to drive Tesla's motor.

He drifted on to Paris soon after and took a job with the new Sociıtı

Continentale Edison, the French Edison company. There he dutifully worked on d.c. lighting systems while inwardly pursuing his own dream. Early in 1883, sent to Strasbourg as a troubleshooter, he collected some spare parts and built the first working model of his motor, and a small dynamo to drive it. Everything worked as he had envisioned it. Tesla was then twenty-seven.

Calling some wealthy industrialists of Strasbourg together, he tried to interest them in forming a company to back him, for he knew that the Edison company had no desire even to hear about an a.c. system. But the Strasbourgers could not understand what Tesla was telling them, and he returned to Paris without the capital he had hoped to raise.

There he came to the attention of Charles Batchelor in the spring of 1884. Batchelor, sensing Tesla's brilliance, invited him to go to the United States and work directly with Edison. Naively confident that he could convert Edison to his own system, Tesla sank his savings into a steamship passage to New York. He arrived with four cents in cash, the clothes on his back, and the plans for his a.c. motor; he also carried a letter of introduction from Batchelor to Edison, saying, "I know two great men and you are one of them; the other is this young man."

Tesla went to Edison's Fifth Avenue office, presented his letter, and was treated to a discourse on the problems of Pearl Street. Edison would hear no talk of alternating current; he put Tesla to work at once on routine electrical work. When Tesla solved a knotty problem with unexpected cleverness, Edison saw his merit and gave him harder assignments—without, however, raising his pay beyond that of a beginning technician. Feeling underpaid, and angered by Edison's refusal to consider his theories, Tesla came to the inevitable collision with his employer and resigned after a year. From the spring of 1886 to the spring of 1887 he supported himself as a day laborer, digging ditches and taking on occasional electrical jobs.

Then he found backers, and set up the Tesla Electric Company. Though their financial support was modest, Tesla was able to carry on his research and to apply for patents. Between November and December, 1887, Tesla filed for seven patents on his motor and on an electrical distribution system; so radically different were they from anything in existence that the applications were granted in only six months. Tesla was invited to explain his ideas at a meeting of the American Institute of Electrical Engineers on May 16, 1888. He had "arrived."

Now he held patents on a revolutionary power system. His high-voltage lines could carry power hundreds of miles with relatively little energy loss or voltage drop. His motors were far more efficient than the small d.c. motors then in use, and could handle much greater loads. But the Tesla Electric Company lacked the capital to make use of the patents. Inevitably, Tesla was summoned to meet George Westinghouse. Westinghouse offered the inventor a million dollars in cash for his patents, plus a generous royalty arrangement, and hired Tesla as a consultant.

Though the unpredictable Tesla soon quarreled with Westinghouse and returned to private research, his patents had passed into the manufacturer's control. Now it was Westinghouse, and not Edison, who held the technological advantage. Westinghouse could make use of the newest and most exciting ideas in the field; Edison, relegated to the role of grim defender of the past, was forced to rely on the great work of 1878-83 at a time when that work had passed into obsolescence. As often happens when a great man must occupy an untenable position, Edison grew more convinced that he was right each time Westinghouse added new proof that he was wrong. Thus began the famed "Battle of the Currents."

Edison set out to prove that a.c. was a menace to the public. He rigged an a.c. generator at West Orange that supplied current at 1,000 volts, and before invited audiences of reporters and other guests staged grisly demonstrations in which hapless dogs and cats were nudged out onto wired sheets of tin and electrocuted. The stray animals of upper New Jersey perished at a ghastly rate in this gruesome promotional campaign.

Then, in February of 1888, Edward Johnson issued a manifesto titled, "A Warning from the Edison Electric Light Company." It summarized the results of the animal executions and provided a long list of the fatalities caused by is lights, which used high voltages. The pamphlet sideswiped Westinghouse and Thomson-Houston by branding them as "patent pirates" who were out to gain quick wealth by introducing a newfangled and risky kind of electricity to the American home.

Westinghouse issued relatively restrained denials, pointing out that the mishaps of arc light companies in 1880 had nothing to do with his own incandescent system of 1888, and that in any event high voltages would be used only in transmission lines; current entering the home would be stepped down to a voltage no greater than that favored by Edison.

Thomson-Houston stayed out of the battle. After its initial delay it had gone into full production of a.c. equipment, which Elihu Thomson at last regarded as perfectly safe. Thomson Houston was selling hundreds of alternators and thousands of transformers now. As Elihu Thomson pointed out, a.c. was actually less deadly than d.c., given the same voltage and amperage.

The only real danger in using a.c. lay in the possibility that a short-circuit in the transformer might allow a high voltage to emerge from the secondary, or step-down coil; but Thomson found a means of protecting against this danger, and had no further fears of trouble. Nevertheless, Charles Coffin, the shrewd head of Thomson-Houston, preferred to let Westinghouse and Edison battle it out in the headlines while his company went ahead with steady output of the new equipment.

Edison's promotional campaign—sparked by the ingenuity of Samuel Insull—swung into a high level in the autumn of 1888. Edison lobbyists asked state legislatures for laws limiting electric circuits to 800 volts, and nearly succeeded in Ohio and Virginia. In New York, Edison took a different and much more bizarre approach: he persuaded the legislature to legalize the electric chair for the execution of condemned criminals—and saw to it that the fatal chair used Westinghouse equipment! What better way to dramatize the lethal nature of alternating current?

Harold P. Brown, a former laboratory assistant at West Orange, served as Edison's chief agent in this enterprise. Brown went before the legislators to preach the virtues of electrical execution—"instantaneous, painless, and humane"—and pushed through a bill, late in 1888, providing for the adoption of the electric chair in place of hanging. The New York State Legislature appointed Brown as a consultant to develop such a chair. With much fanfare he purchased three Westinghouse alternators and put on a display of their potency on March 2, 1889. The victims included several large dogs, four calves, and an elderly horse.

Ten seconds at 800 volts finished off the dogs; the calves were bigger, and had greater resistance, but they died in fifteen seconds. The horse required twenty-five seconds at 1000 volts. For human beings, Brown recommended 2000 volts for rapid and painless dispatch. A human being was available for this research: one William Kemmler, a condemned murderer. The state authorities were eager to try Mr. Edison's death chair, and so Kemmler was sentenced "to suffer death by electricity at Auburn Prison within the week beginning Monday, June 24, 1889."

Kemmler seemed pleased at the attention he would get and the place in history that would be his, but his lawyers objected. Hearings followed; objections were filed. Twice the prisoner was reprieved while more experiments were performed. At one hearing Edison was called to the witness stand to give his evidence. He was interrogated by Deputy Attorney General Poste, who asked, "What is your calling or profession?"

"Inventor," Edison replied.

"Have you devoted a great deal of attention to the subject of electricity?"

"Yes."

"How long have you been engaged in the work of an inventor or electrician?"

"Twenty-six years."

Edison was asked to explain the difference between continuous (direct) and alternating current. "A continuous current," he said, "is one that flows like

like water through a pipe. An alternating current is the same as if a body of water were allowed to flow through the pipe in one direction for a given time and then its direction reversed for a given time."

Had he measured the electrical resistance of human beings? Yes, he said, he had. Measurements of 250 people showed an average resistance of 1000 ohms, the highest being 1800, the lowest 600.

"In your judgment," Poste asked, "can an artificial electric current be generated and applied in such a manner as to produce death in human beings in every case?"

"Yes."

"Instantly?"

"Yes."

W. Bourke Cockran, one of Kemmler's lawyers, now crossexamined the inventor. Cockran was troubled by the point that different men had different resistances. What if Kemmler did not die at once, but met some hideous and lingering doom? "What would be the effect of the current on Kemmler in case the current was applied for five or six minutes?" Cockran asked. "Would he not be carbonized?"

"No," Edison replied. "He would be mummified. All the water in his body would evaporate."

Edison expressed the belief that 1000 volts of one-ampere current would be ten times as much as was needed to kill any man with the Westinghouse alternator.

"That is your belief, not from knowledge?" Cockran asked.

"From belief," said Edison quietly. "I never killed anybody."

After lengthy debate, Kemmler's execution date was set for August 6, 1890. The warden of Auburn Prison was empowered to invite twenty-one witnesses. Most of the guests selected were scientific men, among them Edison—who, however, did not care to be present at Kemmler's death.

An ordinary commercial Westinghouse machine capable of producing a current at 1500 volts was used. It was driven by a steam engine in the prison basement; the power lines were run out of a window of the dynamo room to the roof of the jail and along the roof to the death chamber. A current varying in force from 800 to 1300 volts gave Kemmler his place in the annals of penology; the witnesses unofficially reported that the death had been slower and less pleasant than Edison had predicted.

None the less, electrocution became the rule in New York State, and Edison, at some loss in dignity, managed to frighten many Americans into thinking that a.c. was a public menace. Some of his own men, particularly Frank Sprague, pleaded with him to halt his campaign, but Edison had gone too far now to back down. All his prestige was pledged to direct current. Out of bitterness and testiness he had elevated his original shortsightedness to the status of an inflexible policy; it was the worst mistake of his career.

Through 1888 and 1889 it became increasingly apparent to most men in the field that the future of electricity lay with a.c. By the time William Kemmler went to his death in the summer of 1890, that fact seemed certain to virtually everyone but Edison. Among the converts to the new creed were the executives of Edison General Electric, although they did not dare suggest to Edison that he had outlived his usefulness in electrical engineering. A series of complicated developments grew out of Edison's vested interest in the outmoded system of electrical generation and transmission, and when the dust settled, Edison had been quietly shown to the exit.

Chapter 5

Electrical Circuits: Network Analysis

A network, in the context of electronics, is a collection of interconnected components. Network analysis is the process of finding the voltages across, and the currents through, every component in the network. There are a number of different techniques for achieving this. However, for the most part, they assume that the components of the network are all linear. The methods described in this article are only applicable to *linear* network analysis except where explicitly stated.

Component: A device with two or more terminals into which, or out of which, charge may flow.

Node: A point at which terminals of more than two components are joined. A conductor with a substantially zero resistance is considered to be a node for the purpose of analysis.

Branch: The component(s) joining two nodes.

Mesh: A group of branches within a network joined so as to form a complete loop.

Port: Two terminals where the current into one is identical to the current out of the other.

Circuit: A current from one terminal of a generator, through load component(s) and back into the other terminal. A circuit is, in this sense, a one-port network and is a trivial case to analyse. If there is any connection to any other circuits then a non-trivial network has been formed and at least two ports must exist.

Transfer function: The relationship of the currents and/or voltages between two ports. Most often, an input port and an output port are discussed and the transfer function is described as gain or attenuation.

Component transfer function: For a two-terminal component (i.e. one-port component), the current and voltage are taken as the input and output and the transfer function will have units of impedance or admittance (it is usually a matter of arbitrary convenience whether voltage or current is considered the input). A three (or more) terminal component effectively has two (or more) ports and the transfer function cannot be expressed as a single impedance. The usual approach is to express the transfer function as a matrix of parameters. These parameters can be impedances, but there is a large number of other approaches, see two-port network.

EQUIVALENT CIRCUITS

A useful procedure in network analysis is to simplify the network by reducing the number of components. This can be done by replacing the actual components with other notional components that have the same effect. A particular technique might directly reduce the number of components, for instance by combining impedances in series. On the other hand it might merely change the form in to one in which the components can be reduced in a later operation. For instance, one might transform a voltage generator into a current generator using Norton's theorem in order to be able to later combine the internal resistance of the generator with a parallel impedance load.

A resistive circuit is a circuit containing only resistors, ideal current sources, and ideal voltage sources. If the sources are constant (DC) sources, the result is a DC circuit. The analysis of a circuit refers to the process of solving for the voltages and currents present in the circuit. The solution principles outlined here also apply to phasor analysis of AC circuits.

Two circuits are said to be equivalent with respect to a pair of terminals if the voltage across the terminals and current through the terminals for one network have the same relationship as the voltage and current at the terminals of the other network. If $V_2 = V_1$ implies $I_2 = I_1$ for all (real) values of V_1, then with respect to terminals ab and xy, circuit 1 and circuit 2 are equivalent.

The above is a sufficient definition for a one-port network. For more than one port, then it must be defined that the currents and voltages between all pairs of corresponding ports must bear the same relationship. For instance, star and delta networks are effectively three port networks and hence require three simultaneous equations to fully specify their equivalence.

IMPEDANCES IN SERIES AND IN PARALLEL

Any two terminal network of impedances can eventually be reduced to a single impedance by successive applications of impendances in series or impendances in parallel.

DELTA-WYE TRANSFORMATION

A network of impedances with more than two terminals cannot be reduced to a single impedance equivalent circuit. An n-terminal network can, at best, be reduced to n impedances. For a three terminal network, the three impedances can be expressed as a three node delta (Δ) network or a four node star (Y) network. These two networks are equivalent and the transformations between them are given below. A general network with an arbitrary number of terminals cannot be reduced to the minimum number of impedances using only series and parallel combinations.

In general, Y-Δ and Δ-Y transformations must also be used. It can be shown that this is sufficient to find the minimal network for any arbitrary network with successive applications of series, parallel, Y-Δ and Δ-Y; no more complex transformations are required.

For equivalence, the impedances between any pair of terminals must be the same for both networks, resulting in a set of three simultaneous equations. The equations below are expressed as resistances but apply equally to the general case with impedances. The Y-Δ transform, also written Y-delta, Wye-delta, Kennelly's delta-star transformation, star-mesh transformation, T-Π or T-pi transform, is a mathematical technique to simplify the analysis of an electrical network.

The name derives from the shapes of the circuit diagrams, which look respectively like the letter Y and the Greek capital letter Δ. In the United Kingdom, the wye diagram is sometimes known as a star. This circuit transformation theory was published by Arthur Edwin Kennelly in 1899.

BASIC Y-Δ TRANSFORMATION

The transformation is used to establish equivalence for networks with 3 terminals. Where three elements terminate at a common node and none are sources, the node is eliminated by transforming the impedances. For equivalence, the impedance between any pair of terminals must be the same for both networks. The equations given here are valid for real as well as complex impedances.

GRAPH THEORY

In graph theory, the Y-Δ transform means replacing a Y subgraph of a graph with the equivalent Δ subgraph. The transform preserves the number of edges in a graph, but not the number of vertices or the number of cycles. Two graphs are said to be Y-Δ equivalent if one can be obtained from the other by a series of Y-Δ transforms in either direction. For example, the Petersen graphs are a Y-Δ equivalence class.

DEMONSTRATION

Δ-load to Y-load Transformation Equations

Given the values of R_b, R_c and R_a from the Δ configuration, we want to obtain the values of R_1, R_2 and R_3 in the equivalent Y configuration. In order to do that, we will calculate the equivalent impedances of both configurations in N_1N_2, N_1N_3 and N_2N_3, supposing in each case that the omitted node is unconnected, and we will equal both expressions, since the resistance must be the same.

Source Transformation

A generator with an internal impedance (ie non-ideal generator) can be represented as either an ideal voltage generator or an ideal current generator plus the impedance. These two forms are equivalent and the transformations are given below. If the two networks are equivalent with respect to terminals ab, then V and I must be identical for both networks. Thus,

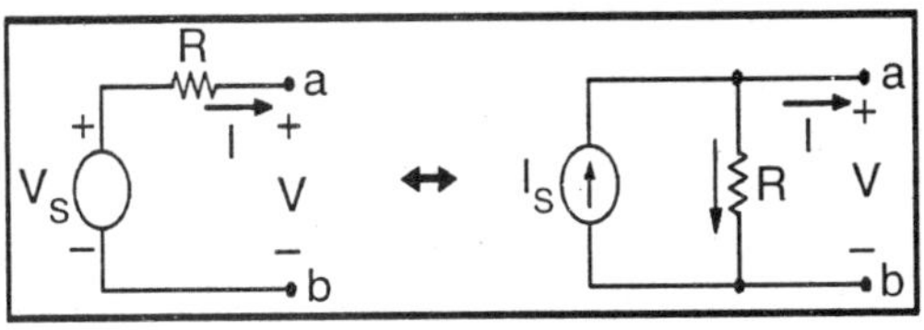

SIMPLE NETWORKS

Some very simple networks can be analysed without the need to apply the more systematic approaches.

VOLTAGE DIVISION OF SERIES COMPONENTS

In electronics, a voltage divider (also known as a potential divider) is a simple linear circuit that produces an output voltage (V_{out}) that is a fraction of its input voltage (V_{in}). Voltage division refers to the partitioning of a voltage among the components of the divider. The formula governing a voltage divider is similar to that for a current divider, but the ratio describing voltage division places the selected impedance in the numerator, unlike current division where it is the unselected components that enter the numerator. A simple example of a voltage divider consists of two resistors in series or a potentiometer. It is commonly used to create a reference voltage, and may also be used as a signal attenuator at low frequencies.

GENERAL CASE

A voltage divider referenced to ground is created by connecting two impedances in series. The input voltage is applied across the series impedances Z_1 and Z_2 and the output is the voltage across Z_2. Z_1 and Z_2 may be composed of any combination of elements such as resistors, inductors and capacitors.

RESISTIVE DIVIDER

A resistive divider is a special case where both impedances, Z_1 and

Z_2, are purely resistive. The ratio then depends on frequency, in this case decreasing as frequency increases. This circuit is, in fact, a basic (first-order) lowpass filter. The ratio contains an imaginary number, and actually contains both the amplitude and phase shift information of the filter.

LOADING EFFECT

The voltage output of a voltage divider is not fixed but varies according to the load. To obtain a reasonably stable output voltage the output current should be a small fraction of the input current. The drawback of this is that most of the input current is wasted as heat in the resistors. The gain of an amplifier generally depends on its source and load terminations, so-called loading effects that reduce the gain.

The analysis of the amplifier itself is conveniently treated separately using idealized drivers and loads, and then supplemented by the use of voltage and current division to include the loading effects of real sources and loads. The choice of idealized driver and idealized load depends upon whether current or voltage is the input/output variable for the amplifier at hand, as described next. For more detail on types of amplifier based upon input/output variables.

In terms of sources, amplifiers with voltage input (voltage and transconductance amplifiers) typically are characterized using ideal zero-impedance voltage sources. In terms of terminations, amplifiers with voltage output (voltage and transresistance amplifiers) typically are characterized in terms of an open circuit output condition. Similarly, amplifiers with current input (current and transresistance amplifiers) are characterized using ideal infinite impedance current sources, while amplifiers with current output (current and transconductance amplifiers) are characterized by a short-circuit output condition,

As stated above, when any of these amplifiers is driven by a non-ideal source, and/or terminated by a finite, non-zero load, the effective gain is lowered due to the loading effect at the input and/or the output. (A current amplifier example is found in the article on current division.) For any of the four types of amplifier (current, voltage, transconductance or transresistance), these loading effects can be understood as a result of voltage division and/or current division, as described next.

INPUT LOADING

A general voltage source can be represented by a Thıvenin equivalent circuit with Thıvenin series impedance R_S. In the same manner, the ideal input current for an ideal driver i_i is realized only for an infinite-resistance current driver. More generally, complex frequency-dependent impedances can be used instead of the driver and amplifier resistances.

OUTPUT LOADING

For a finite load, R_L an output voltage is reduced by voltage division by the factor $R_L/(R_L + R_{out})$, where R_{out} is the amplifier output resistance. Likewise, as the term *short-circuit* implies, the output current delivered to a load R_L is reduced by current division by the factor $R_{out}/(R_L + R_{out})$. The overall gain is reduced below the gain estimated using an ideal load by the same current division factor. More generally, complex frequency-dependent impedances can be used instead of the load and amplifier resistances.

UNILATERAL VERSUS BILATERAL AMPLIFIERS

In a more general case where the amplifier is represented by a two port, the input resistance of the amplifier depends on its load, and the output resistance on the source impedance. The loading factors in these cases must employ the true amplifier impedances including these bilateral effects. That is, the ideal current gain A_i is reduced not only by the loading factors, but due to the bilateral nature of the two-port by an additional factor ($1 + \beta (R_S/R_L) A_{loaded}$), which is typical of negative feedback amplifier circuits. The factor β (R_S/R_L) is the voltage feedback provided by the current feedback source of current gain β A/A. For instance, for an ideal voltage source with $R_S = 0\ \Omega$, the current feedback has no influence, and for $R_L = \infty\ \Omega$, there is zero load current, again disabling the feedback.

REFERENCE VOLTAGE

Voltage dividers are often used to produce stable reference voltages. The term *reference voltage* implies that little or no current is drawn from the divider output node by an attached load. Thus, use of the divider as a reference requires a load device with a high input impedance to avoid loading the divider, that is, to avoid disturbing its output voltage. A simple way of avoiding loading (for low power applications) is to simply input the reference voltage into the non-inverting input of an op-amp buffer.

VOLTAGE SOURCE

While voltage dividers may be used to produce precise reference voltages (that is, when no current is drawn from the reference node), they make poor voltage sources (that is, when current *is* drawn from the reference node). The reason for poor source behaviour is that the current drawn by the load passes through resistor R_1, but not through R_2, causing the voltage drop across R_1 to change with the load current, and thereby changing the output voltage.

In terms of the above equations, if current flows into a load resistance R_L (attached at the output node where the voltage is V_{out}), that load resistance must be considered in parallel with R_2 to determine the voltage at V_{out}. In other words, for high impedance loads it is possible to use a voltage divider as a voltage source, as long as R_2 has very small value compared to the load. This technique leads to considerable power dissipation in the divider.

A voltage divider is commonly used to set the DC bias of a common emitter amplifier, where the current drawn from the divider is the relatively low base current of the transistor.

CURRENT DIVISION OF PARALLEL COMPONENTS

CURRENT DIVISION OF TWO PARALLEL COMPONENTS

Nodal Analysis

- Label all nodes in the circuit. Arbitrarily select any node as reference.
- Define a voltage variable from every remaining node to the reference. These voltage variables must be defined as voltage rises with respect to the reference node.
- Write a KCL equation for every node except the reference.
- Solve the resulting system of equations.

Mesh Analysis

Mesh — a loop that does not contain an inner loop.

- Count the number of "window panes" in the circuit. Assign a mesh current to each window pane.
- Write a KVL equation for every mesh whose current is unknown.
- Solve the resulting equations

SUPERPOSITION

In this method, the effect of each generator in turn is calculated. All the generators other than the one being considered are removed; either short-circuited in the case of voltage generators, or open circuited in the case of current generators. The total current through, or the total voltage across, a particular branch is then calculated by summing all the individual currents or voltages.

There is an underlying assumption to this method that the total current or voltage is a linear superposition of its parts. The method cannot, therefore, be used if non-linear components are present. Note that mesh analysis and node analysis also implicitly use superposition so these too, are only applicable to linear circuits.

Choice of Method

Choice of method is to some extent a matter of taste. If the network is particularly simple or only a specific current or voltage is required then ad-hoc application of some simple equivalent circuits may yield the answer without recourse to the more systematic methods.

- Superposition is possibly the most conceptually simple method but rapidly leads to a large number of equations and messy impedance combinations as the network becomes larger.
- Nodal analysis: The number of voltage variables, and hence simultaneous equations to solve, equals the number of nodes minus one. Every voltage source connected to the reference node reduces the number of unknowns (and equations) by one. Nodal analysis is thus best for voltage sources.
- Mesh analysis: The number of current variables, and hence simultaneous equations to solve, equals the number of meshes. Every current source in a mesh reduces the number of unknowns by one. Mesh analysis is thus best for current sources. Mesh analysis, however, cannot be used with networks which cannot be drawn as a planar network, that is, with no crossing components.

TRANSFER FUNCTION

A transfer function expresses the relationship between an input and an output of a network. For resistive networks, this will always be a simple real number or an expression which boils down to a real number. Resistive networks are represented by a system of simultaneous algebraic equations. However in the general case of linear networks, the network is represented by a system of simultaneous linear differential equations.

In network analysis, rather than use the differential equations directly, it is usual practice to carry out a Laplace transform on them first and then express the result in terms of the Laplace parameter s, which in general is complex. This is described as working in the s-domain. Working with the equations directly would be described as working in the time (or t) domain because the results would be expressed as time varying quantities.

The Laplace transform is the mathematical method of transforming between the s-domain and the t-domain. This approach is standard in control theory and is useful for determining stability of a system, for instance, in an amplifier with feedback.

TWO TERMINAL COMPONENT TRANSFER FUNCTIONS

For two terminal components the transfer function is the relationship between the current input to the device and the resulting voltage across it. The transfer function, Z(s), will thus have units of impedance - ohms.

Two Port Network Transfer Function

Transfer functions, in general, in control theory are given the symbol H(s). Most commonly in electronics, transfer function is defined as the ratio of output voltage to input voltage and given the symbol A(s), or more commonly (because analysis is invariably done in terms of sine wave response), A(jω).

The A standing for attenuation, or amplification, depending on context. In general, this will be a complex function of $j\omega$, which can be derived from an analysis of the impedances in the network and their individual transfer functions. Sometimes the analyst is only interested in the magnitiude of the gain and not the phase angle.

Two Port Parameters

The concept of a two-port network can be useful in network analysis as a black box approach to analysis. The behaviour of the two-port network in a larger network can be entirely characterised without necessarily stating anything about the internal structure. However, to do this it is necessary to have more information than just the A(jω) described above. It can be shown that four such parameters are required to fully characterise the two-port network.

These could be the forward transfer function, the input impedance, the reverse transfer function (ie, the voltage appearing at the input when a voltage is applied to the output) and the output impedance. There are many others, one of these expresses all four parameters as impedances. These concepts are capable of being extended to networks of more than two ports. However, this is rarely done in reality as in many practical cases ports are considered either purely input or purely output. If reverse direction transfer functions are ignored, a multi-port network can always be decomposed into a number of two-port networks.

Distributed Components

Where a network is composed of discrete components, analysis using two-port networks is a matter of choice, not essential. The network can always alternatively be analysed in terms of its individual component transfer functions. However, if a network contains distributed components, such as in the case of a transmission line, then it is not possible to analyse in terms of individual components since they do not exist.

The most common approach to this is to model the line as a two-port network and characterise it using two-port parameters (or something equivalent to them). Another example of this technique is modelling the carriers crossing the base region in a high frequency transistor. The base region has to be modelled as distributed resistance and capacitance rather than lumped components.

Image Analysis

Transmission lines and certain types of filter design use the image method to determine their transfer parameters. In this method, the behaviour of an infinitely long cascade connected chain of identical networks is considered. The input and output impedances and the forward and reverse transmission functions are then calculated for this infinitely long chain.

Although, the theoretical values so obtained can never be exactly realised in practice, in many cases they serve as a very good approximation for the behaviour of a finite chain as long as it is not too short.

Non-linear Networks

Most electronic designs are, in reality, non-linear. There is very little that does not include some semiconductor devices. There are many other ways that non-linearity can appear in a network. All methods utilising linear superposition will fail when non-linear components are present. There are several options for dealing with non-linearity depending on the type of circuit and the information the analyst wishes to obtain.

BOOLEAN ANALYSIS OF SWITCHING NETWORKS

A switching device is one where the non-linearity is utilised to produce two opposite states. CMOS devices in digital circuits, for instance, have their output connected to either the positive or the negative supply rail and are never found at anything in between except during a transient period when the device is actually switching.

Here the non-linearity is designed to be extreme, and the analyst can actually take advantage of that fact. These kinds of networks can be analysed using Boolean algebra by assigning the two states ("on"/"off", "positive"/"negative" or whatever states are being used) to the boolean constants "0" and "1".

The transients are ignored in this analysis, along with any slight discrepancy between the actual state of the device and the nominal state assigned to a boolean value. For instance, boolean "1" may be assigned to the state of +5V. The output of the device may actually be +4.5V but the analyst still considers this to be boolean "1". Device manufacturers will usually specify a range of values in their data sheets that are to be considered undefined (ie the result will be unpredictable).

The transients are not entirely uninteresting to the analyst. The maximum rate of switching is determined by the speed of transition from one state to the other. Happily for the analyst, for many devices most of the transition occurs in the linear portion of the devices transfer function and linear analysis can be applied to obtain at least an approximate answer. It is mathematically possible to derive boolean algebras which have more than two states. There is not too much use found for these in electronics, although three-state devices are passingly common.

Separation of Bias and Signal Analyses

This technique is used where the operation of the circuit is to be essentially linear, but the devices used to implement it are non-linear. A transistor amplifier is an example of this kind of network. The essence of this technique is to separate the analysis in to two parts. Firstly, the dc biases are analysed using some non-linear method. This establishes the quiescent operating point of the circuit. Secondly, the small signal characteristics of the circuit are analysed using linear network analysis. Examples of methods that can be used for both these stages are given below.

Graphical Method of dc Analysis

In a great many circuit designs, the dc bias is fed to a non-linear component via a resistor (or possibly a network of resistors). Since resistors are linear components, it is particularly easy to determine the quiescent operating point of the non-linear device from a graph of its transfer function. The method is as follows: from linear network analysis the output transfer function (that is output voltage against output current) is calculated for the network of resistor(s) and the generator driving them. This will be a straight line and can readily be superimposed on the transfer function plot of the non-linear device. The point where the lines cross is the quiescent operating point.

Perhaps the easiest practical method is to calculate the (linear) network open circuit voltage and short circuit current and plot these on the transfer function of the non-linear device. The straight line joining these two point is the transfer function of the network. In reality, the designer of the circuit would proceed in the reverse direction to that described. Starting from a plot provided in the manufacturers data sheet for the non-linear device, the designer would choose the desired operating point and then calculate the linear component values required to achieve it.

It is still possible to use this method if the device being biased has its bias fed through another device which is itself non-linear - a diode for instance. In this case however, the plot of the network transfer function onto the device being biased would no longer be a straight line and is consequently more tedious to do.

SMALL SIGNAL EQUIVALENT CIRCUIT

This method can be used where the deviation of the input and output signals in a network stay within a substantially linear portion of the non-linear devices transfer function, or else are so small that the curve of the transfer function can be considered linear. Under a set of these specific conditions, the non-linear device can be represented by an equivalent linear network. It must be remembered that this equivalent circuit is entirely notional and only valid for the small signal deviations. It is entirely inapplicable to the dc biasing of the device.

For a simple two-terminal device, the small signal equivalent circuit may be no more than two components. A resistance equal to the slope of the v/i curve at the operating point (called the dynamic resistance), and tangent to the curve. A generator, because this tangent will not, in general, pass through the origin. With more terminals, more complicated equivalent circuits are required.

A popular form of specifying the small signal equivalent circuit amongst transistor manufacturers is to use the two-port network parameters known as [h] parameters. These are a matrix of four parameters as with the [z] parameters but in the case of the [h] parameters they are a hybrid mixture of impedances, admittances, current gains and voltage gains. In this model the three terminal transistor is considered to be a two port network, one of its terminals being common to both ports. The [h] parameters are quite different depending on which terminal is chosen as the common one. The most important parameter for transistors is usually the forward current gain, h_{21}, in the common emitter configuration. This is designated h_{fe} on data sheets.

The small signal equivalent circuit in terms of two-port parameters leads to the concept of dependent generators. That is, the value of a voltage or current generator depends linearly on a voltage or current elsewhere in the circuit.

There will always be dependent generators in a two-port parameter equivalent circuit. This applies to the [h] parameters as well as to the [z] and any other kind. These dependencies must be preserved when developing the equations in a larger linear network analysis.

PIECEWISE LINEAR METHOD

In this method, the transfer function of the non-linear device is broken up into regions. Each of these regions is approximated by a straight line. Thus, the transfer function will be linear up to a particular point where there will be a discontinuity. Past this point the transfer function will again be linear but with a different slope.

A well known application of this method is the approximation of the transfer function of a pn junction diode. The actual transfer function of an ideal diode has been given at the top of this (non-linear) section. However, this formula is rarely used in network analysis, a piecewise approximation being used instead. It can be seen that the diode current rapidly diminishes to $-I_0$ as the voltage falls. This current, for most purposes, is so small it can be ignored. With increasing voltage, the current increases exponentially. The diode is modelled as an open circuit up to the knee of the exponential curve, then past this point as a resistor equal to the bulk resistance of the semiconducting material.

The commonly accepted values for the transition point voltage are 0.7V for silicon devices and 0.3V for germanium devices. An even simpler model of the diode, sometimes used in switching applications, is short circuit for forward voltages and open circuit for reverse voltages. The model of a forward biased pn junction having an approximately constant 0.7V is also a much used approximation for transistor base-emitter junction voltage in amplifier design.

The piecewise method is similar to the small signal method in that linear network analysis techniques can only be applied if the signal stays within certain bounds. If the signal crosses a discontinuity point then the model is no longer valid for linear analysis purposes. The model does have the advantage over small signal however, in that it is equally applicable to signal and dc bias. These can therefore both be analysed in the same operations and will be linearly superimposable.

Chapter 6

Circuit Theory

Circuit theory is the theory of accomplishing work by means of routing matter through a loop. The types of matter used are:

- In electronic or electrical circuits: electrons (and charged ions, both positive and negative)
- In pneumatic circuits: compressed gas (normally ordinary air)
- In hydraulic circuits: pressurized, relatively incompressible fluid

PARTS OF A CIRCUIT

Every circuit consists of three basic components:

- "Active components": Source of energy
- Transmission lines
  - Control devices (optional)
- "Passive components": Load

A gun, a rocket and an internal combustion engine all use compressed gas to do work, but the spent gas is vented to the atmosphere and is not reused in the system, so these are not examples of pneumatic circuits. Refrigeration systems do, however, recycle the compressed gases they use, but are not typically thought of as circuits. Gears, levers, linkages, pulleys/ropes and sprockets/chains transmit work energy from one location to another, but there is no loop, so these are not examples of circuits.

CIRCUIT VS. NETWORK

An electrical circuit is a collection of electrical components which accomplish a specific task such as heating, lighting or running a motor. This collection may or may not form a complete topological loop, depending on whether it is presently connected to power, integrated into a larger device or circuit, or damaged. Sometimes, it is convenient to speak of an electrical circuit as a network, de-emphasizing the return path. Return paths are sometimes omitted from circuit diagrams, making the resulting graphic visually resemble a network topology rather than some sort of loop topology.

OPEN CIRCUIT VS. CLOSED CIRCUIT

A fundamental part of circuit analysis is determining whether the matter has a return path to the power source. If the matter is blocked from returning to the power source, either wholly or partially, the entire assemblage will be prevented from accomplishing work. In an electrical circuit, an open circuit is caused intentionally when a user opens a switch or unintentionally when vibration or mechanical damage severs a wire. In a pneumatic or hydraulic circuit, this occurs when a valve is closed or there is a leak in one of the lines or components.

In electrical circuits, closing a switch creates a closed loop for the electrons to flow through. This is sometimes referred to as "completing the circuit." Other synonyms are also used.

SHORT CIRCUIT

In an electrical or electronic circuit, sometimes an unintended connection is made, such as when insulation is broken, frayed, melted or chewed by rodents, or a technician inserts a metal tool into a live device. When this happens, current bypasses some or all of the components in the circuit, taking a "shorter" path back to the power source. This can lead to excessive current drain, which in turn generates excessive heat, damaging or destroying sensitive parts of the system such as transistors and ICs.

LOOPS

In Graph theory, an edge whose two ends meet is called a loop, which is an entirely different usage of the word. In any kind of circuit, such a loop has no distinct function. An argument can be made that redundant lines for transmission of power do have a function, even if it is only a backup function.

Types

There are three basic types of circuit currently used in industry:

- Electronic or electrical
- Pneumatic
- Hydraulic

The following is a rough list of the types of components which make up each type of circuit.

Electronic Circuit

- Sources of energy
  - Batteries
  - Generators
  - Solar cells
- Transmission lines

  - Wires
  - Switches
- Passive components
  - Transducers

Pneumatic Circuit

- Sources of energy
  - Compressor
- Transmission lines
  - Air tank
  - Pneumatic hoses
  - Open atmosphere (for returning the spent gas to the compressor)
  - Valves
- Passive components
  - Pneumatic cylinders

Hydraulic Circuit

- Sources of energy
  - Power pack
- Transmission lines
  - Hydraulic hoses
- Passive components
  - Hydraulic cylinders

SIGNAL (ELECTRICAL ENGINEERING)

In the fields of communications, signal processing, and in electrical engineering more generally, a signal is any time-varying or spatial-varying quantity. In the physical world, any quantity measurable through time or over space can be taken as a signal. Within a complex society, any set of human information or machine data can also be taken as a signal.

Such information or machine data (for example, the dots on a screen, the ink making up text on a paper page, or the words now flowing into the reader's mind) must all be part of systems existing in the physical world – either living or non-living. Despite the complexity and even mystery – in the case of the reader's mind – of such systems, their outputs and inputs can often be represented with great fidelity as simple quantities measurable through time or across space.

In the latter half of the 20th century, electrical engineering itself separated into several disciplines, specializing in the design and analysis of physical signals and systems, on the one hand, and in the functional behaviour and conceptual structure of the complex human and machine systems, on the other. These engineering disciplines have led the way in the design, study, and implementation of systems that take advantage of signals as simple measurable quantities in order to facilitate the transmission, storage, and manipulation of information.

SOME DEFINITIONS

Definitions specific to subfields are common. For example, in information theory, a *signal* is a codified message, that is, the sequence of states in a communication channel that encodes a message. In a *communication system*, a *transmitter* encodes a *message* into a signal, which is carried to a *receiver* by the communications *channel*. For example, the words "Mary had a little lamb" might be the message spoken into a telephone. The telephone transmitter converts the sounds into an electrical voltage signal.

The signal is transmitted to the receiving telephone by wires; and at the receiver it is reconverted into sounds. Signals can be categorized in various ways. The most common distinction is between discrete and continuous spaces that the functions are defined over, for example discrete and continuous time domains.

Discrete-time signals are often referred to as *time series* in other fields. Continuous-time signals are often referred to as *continuous signals* even when the signal functions are not continuous; an example is a square-wave signal. A second important distinction is between discrete-valued and continuous-valued. Digital signals are discrete-valued, but are often derived from an underlying continuous-valued physical process.

Discrete-time and Continuous-time Signals

If for a signal, the quantities are defined only on a discrete set of times, we call it a discrete-time signal. In other words, a discrete-time real (or complex) signal can be seen as a function from the set of integers to the set of real (or complex) numbers. A continuous-time real (or complex) signal is any real-valued (or complex-valued) function which is defined for all time *t* in an interval, most commonly an infinite interval.

Analog and Digital Signals

Less formally than the theoretical distinctions mentioned above, two main types of signals encountered in practice are *analog* and *digital*. In short, the difference between them is that digital signals are *discrete* and *quantized*, as defined below, while analog signals possess neither property.

DISCRETIZATION

One of the fundamental distinctions between different types of signals is between continuous and discrete time. In the mathematical abstraction, the domain of a continuous-time (CT) signal is the set of real numbers (or some interval thereof), whereas the domain of a discrete-time (DT) signal is the set of integers (or some interval). What these integers represent depends on the nature of the signal.

DT signals often arise via sampling of CT signals. An audio signal, for example consists of a continually fluxuating voltage on a line that can be digitized by an ADC circuit, wherein the circuit will read the voltage level on the line, say, every 50 μs. The resulting stream of numbers is stored as digital data on a discrete-time signal. Computers and other digital devices are restricted to discrete time.

A discrete signal or discrete-time signal is a time series, perhaps a signal that has been sampled from a continuous-time signal. Unlike a continuous-time signal, a discrete-time signal is not a function of a continuous-time argument, but is a sequence of quantities; that is, a function over a domain of discrete integers. Each value in the sequence is called a sample.

When a discrete-time signal is a sequence corresponding to uniformly spaced times, it has an associated sampling rate; the sampling rate is not apparent in the data sequence, so may be associated as a separate data item.

DIGITAL SIGNALS

A digital signal is a discrete-time signal that takes on only a discrete set of values. It typically derives from a discrete signal that has been quantized.

Common practical digital signals are represented as 8-bit (256 levels), 16-bit (65,536 levels), 32-bit (4.3 billion levels), and so on, though any number of quantization levels is possible, not just powers of two.

QUANTIZATION

If a signal is to be represented as a sequence of numbers, it is impossible to maintain arbitrarily high precision - each number in the sequence must have a finite number of digits. As a result, the values of such a signal are restricted to belong to a finite set; in other words, it is quantized. In digital signal processing, quantization is the process of approximating a continuous range of values (or a very large set of possible discrete values) by a relatively small set of discrete symbols or integer values.

More specifically, a signal can be multi-dimensional and quantization need not be applied to all dimensions. Discrete signals (a common mathematical model) need not be quantized, which can be a point of confusion. A common use of quantization is in the conversion of a discrete signal (a sampled continuous signal) into a digital signal by quantizing.

Both of these steps (sampling and quantizing) are performed in analog-to-digital converters with the quantization level specified in bits.

A specific example would be compact disc (CD) audio which is sampled at 44,100 Hz and quantized with 16 bits (2 bytes) which can be one of 65,536 (i.e. 2) possible values per sample. In electronics, adaptive quantization is a quantization process that varies the step size based on the changes of the input signal, as a means of efficient compression.Two approaches commonly used are forward adaptive quantization and backward adaptive quantization.

Mathematical Description

The simplest and best-known form of quantization is referred to as scalar quantization, since it operates on scalar (as opposed to multi-dimensional vector) input data. The integer-valued quantization index *i* is the representation that is typically stored or transmitted, and then the final interpretation is constructed using $g(i)$ when the data is later interpreted.

In computer audio and most other applications, a method known as *uniform quantization* is the most common. There are two common variations of uniform quantization, called *mid-rise* and *mid-tread* uniform quantizers.

In this case the $f(x)$ and $g(i)$ operators are just multiplying scale factors (one multiplier being the inverse of the other) along with an offset in $g(i)$ function to place the representation value in the middle of the input region for each quantization index.

The value $2^{-(M-)}$ is often referred to as the *quantization step size*. Using this quantization law and assuming that quantization noise is approximately uniformly distributed over the quantization step size (an assumption typically accurate for rapidly varying *x* or high *M*) and further assuming that the input signal *x* to be quantized is approximately uniformly distributed over the entire interval from -1 to 1, the signal to noise ratio (SNR) of the quantization.

From this equation, it is often said that the SNR is approximately 6 dB per bit. For mid-tread uniform quantization, the offset of 0.5 would be added within the floor function instead of outside of it. Sometimes, mid-rise quantization is used without adding the offset of 0.5. This reduces the signal to noise ratio by approximately 6.02 dB, but may be acceptable for the sake of simplicity when the step size is small.

In digital telephony, two popular quantization schemes are the 'A-law' (dominant in Europe) and 'μ-law' (dominant in North America and Japan). These schemes map discrete analog values to an 8-bit scale that is nearly linear for small values and then increases logarithmically as amplitude grows. Because the human ear's perception of loudness is roughly logarithmic, this provides a higher signal to noise ratio over the range of audible sound intensities for a given number of bits.

QUANTIZATION AND DATA COMPRESSION

Quantization plays a major part in lossy data compression. In many cases, quantization can be viewed as the fundamental element that distinguishes lossy data compression from lossless data compression, and the use of quantization is nearly always motivated by the need to reduce the amount of data needed to represent a signal. In some compression schemes, like MP3 or Vorbis, compression is also achieved by selectively discarding some data, an action that can be analyzed as a quantization process (e.g., a vector quantization process) or can be considered a different kind of lossy process.

One example of a lossy compression scheme that uses quantization is JPEG image compression. During JPEG encoding, the data representing an image (typically 8-bits for each of three colour components per pixel) is processed using a discrete cosine transform and is then quantized and entropy coded. By reducing the precision of the transformed values using quantization, the number of bits needed to represent the image can be reduced substantially. For example, images can often be represented with acceptable quality using JPEG at less than 3 bits per pixel (as opposed to the typical 24 bits per pixel needed prior to JPEG compression). Even the original representation using 24 bits per pixel requires quantization for its PCM sampling structure.

In modern compression technology, the entropy of the output of a quantizer matters more than the number of possible values of its output (the number of values being 2^M in the above example). In order to determine how many bits are necessary to effect a given precision, algorithms are used. Suppose, for example, that it is necessary to record six significant digits, that is to say, millionths. The number of values that can be expressed by N bits is equal to two to the Nth power.

To express six decimal digits, the required number of bits is determined by rounding (6/log 2)—where log refers to the base ten, or common, logarithm—up to the nearest integer. Since the logarithm of 2, base ten, is approximately 0.30102, the required number of bits is then given by (6/ 0.30102), or 19.932, rounded up to the nearest integer, *viz.*, 20 bits. This type of quantization—where a set of binary digits, *e.g.*, an arithmetic register in a CPU, are used to represent a quantity—is called Vernier quantization. It is also possible, although rather less efficient, to rely upon equally spaced quantization levels.

This is only practical when a small range of values is expected to be captured: for example, a set of eight possible values requires eight equally spaced quantization levels—which is not unreasonable, although obviously less efficient than a mere trio of binary digits (bits)—but a set of, say, sixty-four possible values, requiring sixty-four equally spaced quantization levels, can be expressed using only six bits, which is obviously far more efficient.

RELATION TO QUANTIZATION IN NATURE

At the most fundamental level, some physical quantities are quantized. This is a result of quantum mechanics). Signals may be treated as continuous for mathematical simplicity by considering the small quantizations as negligible. In any practical application, this inherent quantization is irrelevant for two reasons. First, it is overshadowed by signal noise, the intrusion of extraneous phenomena present in the system upon the signal of interest.

The second, which appears only in measurement applications, is the inaccuracy of instruments. Thus, although all physical signals are intrinsically quantized, the error introduced by modeling them as continuous is vanishingly small.

Examples of Signals

- *Motion.* The motion of a particle through some space can be considered to be a signal, or can be represented by a signal. The domain of a motion signal is one-dimensional (time), and the range is generally three-dimensional. Position is thus a 3-vector signal; position and orientation is a 6-vector signal.
- *Sound.* Since a sound is a vibration of a medium (such as air), a sound signal associates a pressure value to every value of time and three space coordinates. A microphone converts sound pressure at some place to just a function of time, using a voltage signal as an analog of the sound signal.
- *Compact discs* (CDs). CDs contain discrete signals representing sound, recorded at 44,100 samples per second. Each sample contains data for a left and right channel, which may be considered to be a 2-vector (since CDs are recorded in stereo).
- *Pictures.* A picture assigns a colour value to each of a set of points. Since the points lie on a plane, the domain is two-dimensional. If the picture is a physical object, such as a painting, it's a continuous signal. If the picture is a digital image, it's a discrete signal. It's often convenient to represent colour as the sum of the intensities of three primary colours, so that the signal is vector-valued with dimension three.
- *Videos.* A video signal is a sequence of images. A point in a video is identified by its position (two-dimensional) and by the

time at which it occurs, so a video signal has a three-dimensional domain. Analog video has one continuous domain dimension (across a scan line) and two discrete dimensions (frame and line).

- *Biological membrane potentials.* The value of the signal is a straightforward electric potential ("voltage"). The domain is more difficult to establish. Some cells or organelles have the same membrane potential throughout; neurons generally have different potentials at different points. These signals have very low energies, but are enough to make nervous systems work; they can be measured in aggregate by the techniques of electrophysiology.

FREQUENCY ANALYSIS

Signals are often analyzed or modeled in terms of their frequency spectrum. Frequency domain techniques are applicable to all signals, both continuous-time and discrete-time. If a signal is passed through an LTI system, the frequency spectrum of the resulting output signal is the product of the frequency spectrum of the original input signal and the frequency response of the system.

Frequency domain is a term used to describe the analysis of mathematical functions or signals with respect to frequency. Speaking non-technically, a time-domain graph shows how a signal changes over time, whereas a frequency-domain graph shows how much of the signal lies within each given frequency band over a range of frequencies.

A frequency-domain representation can also include information on the phase shift that must be applied to each sinusoid in order to be able to recombine the frequency components to recover the original time signal. The frequency domain relates to the Fourier transform or Fourier series by decomposing a function into an infinite or finite number of frequencies.

This is based on the concept of Fourier series that any waveform can be expressed as a sum of sinusoids (sometimes infinitely many.) A spectrum analyzer is the tool commonly used to visualize real-world signals in the frequency domain.

MAGNITUDE AND PHASE

In using the Laplace, Z-, or Fourier transforms, the frequency spectrum is complex, describing the magnitude and phase of a signal, or of the response of a system, as a function of frequency. In many applications, phase information is not important. By discarding the phase information it is possible to simplify the information in a frequency domain representation to generate a frequency spectrum or spectral density.

A spectrum analyzer is a device that displays the spectrum. The power spectral density is a frequency-domain description that can be applied to a large class of signals that are neither periodic nor square-integrable; to have a power spectral density a signal needs only to be the output of a wide-sense stationary random process.

PARTIAL FREQUENCY-DOMAIN EXAMPLE

Due to popular simplifications of the hearing process and titles such as Plomp's "The Ear as a Frequency Analyzer," the inner ear is often thought of as converting time-domain sound waveforms to frequency-domain spectra. The frequency domain is not actually a very accurate or useful model for hearing, but a time/frequency space or time/place space can be a useful description.

BRIDGE CIRCUIT

A bridge circuit is a type of electrical circuit in which the current in a conductor splits into two parallel paths and then recombines into a single conductor, thereby enclosing a loop. It was originally used for measurement purposes, but can also be used in power supplies.

The best-known bridge circuit, the Wheatstone bridge, was invented by Samuel Hunter Christie and popularized by Charles Wheatstone, and is used for measuring resistance. It is constructed from four resistors, one of which has an unknown value (R_x), one of which is variable (R_2), and two of which are fixed and equal (R_1 and R_3), connected as the sides of a square. Two opposite corners of the square are connected to a source of electrical current, such as a battery. A galvanometer is connected across the other two opposite corners. The variable resistor is adjusted until the galvanometer reads zero.

It is then known that the ratio between the variable resistor and its neighbour is equal to the ratio between the unknown resistor and its neighbour, and this enables the value of the unknown resistor to be calculated. The Wheatstone bridge has also been developed to measure impedance in AC circuits, resulting in designs such as the Wien bridge, the Maxwell bridge and the Heaviside bridge. All are based on the same principle, which is to compare the output of two potentiometers sharing a common source.

In power supply design, a bridge circuit or bridge rectifier is an arrangement of diodes or similar devices used to rectify an electric current, i.e. to convert it from an unknown or alternating polarity to a direct current of known polarity. In some motor controllers, a H-bridge is used to control the direction the motor turns.

DIGITAL ELECTRONICS

Digital electronics are electronics systems that use digital signals.

Digital electronics are representations of Boolean algebra and are used in computers, mobile phones, and other consumer products. In a digital circuit, a signal is represented in one of two states or logic levels. The advantages of digital techniques stem from the fact it is easier to get an electronic device to switch into one of two states, than to accurately reproduce a continuous range of values.

Digital electronics or any digital circuit are usually made from large assemblies of logic gates, simple electronic representations of Boolean logic functions. To most electronic engineers, the terms "digital circuit", "digital system" and "logic" are interchangeable in the context of digital circuits.

ADVANTAGES

One advantage of digital circuits when compared to analog circuits is that signals represented digitally can be transmitted without degrading because of noise. For example, a continuous audio signal, transmitted as a sequence of 1s and 0s, can be reconstructed without error provided the noise picked up in transmission is not enough to prevent identification of the 1s and 0s. An hour of music can be stored on a compact disc as about 6 billion binary digits.

In a digital system, a more precise representation of a signal can be obtained by using more binary digits to represent it. While this requires more digital circuits to process the signals, each digit is handled by the same kind of hardware. In an analog system, additional resolution requires fundamental improvements in the linearity and noise charactersitics of each step of the signal chain. Computer-controlled digital systems can be controlled by software, allowing new functions to be added without changing hardware.

Often this can be done outside of the factory by updating the product's software. So, the product's design errors can be corrected after the product is in a customer's hands. Information storage can be easier in digital systems than in analog ones. The noise-immunity of digital systems permits data to be stored and retrieved without degradation. In an analog system, noise from aging and wear degrade the information stored. In a digital system, as long as the total noise is below a certain level, the information can be recovered perfectly.

DISADVANTAGES

In some cases, digital circuits use more energy than analog circuits to accomplish the same tasks, thus producing more heat as well. In portable or battery-powered systems this can limit use of digital systems. For example, battery-powered cellular telephones often use a low-power analog front-end to amplify and tune in the radio signals from the base station.

However, a base station has grid power and can use power-hungry, but very flexible software radios. Such base stations can be easily reprogrammed to process the signals used in new cellular standards. Digital circuits are sometimes more expensive, especially in small quantities. The sensed world is analog, and signals from this world are analog quantities.

For example, light, temperature, sound, electrical conductivity, electric and magnetic fields are analog. Most useful digital systems must translate from continuous analog signals to discrete digital signals. This causes quantization errors. Quantization error can be reduced if the system stores enough digital data to represent the signal to the desired degree of fidelity. The Nyquist-Shannon sampling theorem provides an important guideline as to how much digital data is needed to accurately portray a given analog signal.

In some systems, if a single piece of digital data is lost or misinterpreted, the meaning of large blocks of related data can completely change. Because of the cliff effect, it can be difficult for users to tell if a particular system is right on the edge of failure, or if it can tolerate much more noise before failing. Digital fragility can be reduced by designing a digital system for robustness. For example, a parity bit or other error management method can be inserted into the signal path.

These schemes help the system detect errors, and then either correct the errors, or at least ask for a new copy of the data. In a state-machine, the state transition logic can be designed to catch unused states and trigger a reset sequence or other error recovery routine. Embedded software designs that employ Immunity Aware Programming, such as the practice of filling unused programme memory with interrupt instructions that point to an error recovery routine. This helps guard against failures that corrupt the microcontroller's instruction pointer which could otherwise cause random code to be executed. Digital memory and transmission systems can use techniques such as error detection and correction to use additional data to correct any errors in transmission and storage. On the other hand, some techniques used in digital systems make those systems more vulnerable to single-bit errors. These techniques are acceptable when the underlying bits are reliable enough that such errors are highly unlikely.

- A single-bit error in audio data stored directly as linear pulse code modulation (such as on a CD-ROM causes, at worst, a single click). Instead, many people use audio compression to save storage space and download time, even though a single-bit error may corrupt the entire song.

ANALOG ISSUES IN DIGITAL CIRCUITS

Digital circuits are made from analog components. The design must assure that the analog nature of the components doesn't dominate the desired digital behaviour. Digital systems must manage noise and timing margins, parasitic inductances and capacitances, and filter power connections.

Bad designs have intermittent problems such as “glitches”, vanishingly-fast pulses that may trigger some logic but not others, “runt pulses” that do not reach valid “threshold” voltages, or unexpected (“undecoded”) combinations of logic states. Since digital circuits are made from analog components, digital circuits calculate more slowly than low-precision analog circuits that use a similar amount of space and power.

However, the digital circuit will calculate more repeatably, because of its high noise immunity. On the other hand, in the high-precision domain (for example, where 14 or more bits of precision are needed), analog circuits require much more power and area than digital equivalents.

Construction

A digital circuit is often constructed from small electronic circuits called logic gates. Each logic gate represents a function of boolean logic. A logic gate is an arrangement of electrically controlled switches. The output of a logic gate is an electrical flow or voltage, that can, in turn, control more logic gates.

Logic gates often use the fewest number of transistors in order to reduce their size, power consumption and cost, and increase their reliability. Integrated circuits, are the least expensive way to make logic gates in large volumes. Integrated circuits are usually designed by engineers using electronic design automation software.

Another form of digital circuit is constructed from lookup tables, (many sold as “programmable logic devices”, though other kinds of PLDs exist). Lookup tables can perform the same functions as machines based on logic gates, but can be easily reprogrammed without changing the wiring. This means that a designer can often repair design errors without changing the arrangement of wires. Therefore, in small volume products, programmable logic devices are often the preferred solution. They are usually designed by engineers using electronic design automation software.

When the volumes are medium to large, and the logic can be slow, or involves complex algorithms or sequences, often a small microcontroller is programmed to make an embedded system. These are usually programmed by software engineers. When only one digital circuit is needed, and its design is totally customized, as for a factory production line controller, the conventional solution is a programmable logic controller, or PLC. These are usually programmed by electricians, using ladder logic.

STRUCTURE OF DIGITAL SYSTEMS

Engineers use many methods to minimize logic functions, in order to reduce the circuit's complexity. When the complexity is less, the circuit also has fewer errors and less electronics, and is therefore less expensive. The most widely used simplification is a minimization algorithm like the Espresso heuristic logic minimizer within a CAD system, although historically, binary decision diagrams, an automated Quine–McCluskey algorithm, truth tables, Karnaugh Maps, and Boolean algebra have been used.

Representations are crucial to an engineer's design of digital circuits. Some analysis methods only work with particular representations. The classical way to represent a digital circuit is with an equivalent set of logic gates. Another way, often with the least electronics, is to construct an equivalent system of electronic switches (usually transistors). One of the easiest ways is to simply have a memory containing a truth table. The inputs are fed into the address of the memory, and the data outputs of the memory become the outputs.

For automated analysis, these representations have digital file formats that can be processed by computer programs. Most digital engineers are very careful to select computer programs ("tools") with compatible file formats. To choose representations, engineers consider types of digital systems. Most digital systems divide into "combinatorial systems" and "sequential systems." A combinatorial system always presents the same output when given the same inputs. It is basically a representation of a set of logic functions, as already discussed.

A sequential system is a combinatorial system with some of the outputs fed back as inputs. This makes the digital machine perform a "sequence" of operations. The simplest sequential system is probably a flip flop, a mechanism that represents a binary digit or "bit". Sequential systems are often designed as state machines. In this way, engineers can design a system's gross behaviour, and even test it in a simulation, without considering all the details of the logic functions.

Sequential systems divide into two further subcategories. "Synchronous" sequential systems change state all at once, when a "clock" signal changes state. "Asynchronous" sequential systems propagate changes whenever inputs change. Synchronous sequential systems are made of well-characterized asynchronous circuits such as flip-flops, that change only when the clock changes, and which have carefully designed timing margins.

The usual way to implement a synchronous sequential state machine is divide it into a piece of combinatorial logic and a set of flip flops called a "state register." Each time a clock signal ticks, the state register captures the feedback generated from the previous state of the combinatorial logic, and feeds it back as an unchanging input to the combinatorial part of the state machine. The fastest rate of the clock is set by the most time-consuming logic calculation in the combinatorial logic.

The state register is just a representation of a binary number. If the states in the state machine are numbered (easy to arrange), the logic function is some combinatorial logic that produces the number of the next state. In comparison, asynchronous systems are very hard to design because all possible states, in all possible timings must be considered. The usual method is to construct a table of the minimum and maximum time that each such state can exist, and then adjust the circuit to minimize the number of such states, and force the circuit to periodically wait for all of its parts to enter a compatible state. (This is called "self-resynchronization.")

Without such careful design, it is easy to accidentally produce asynchronous logic that is "unstable", that is, real electronics will have unpredictable results because of the cumulative delays caused by small variations in the values of the electronic components. Certain circuits (such as the synchronizer flip-flops, switch debouncers, and the like which allow external unsynchronized signals to enter synchronous logic circuits) are inherently asynchronous in their design and must be analyzed as such.

As of 2005, almost all digital machines are synchronous designs because it is much easier to create and verify a synchronous design — the software currently used to simulate digital machines does not yet handle asynchronous designs. However, asynchronous logic is thought to be superior, if it can be made to work, because its speed is not constrained by an arbitrary clock; instead, it simply runs at the maximum speed permitted by the propagation rates of the logic gates from which it is constructed. Building an asynchronous circuit using faster parts implicitly makes the circuit "go" faster.

More generally, many digital systems are data flow machines. These are usually designed using synchronous register transfer logic, using hardware description languages such as VHDL or Verilog. In register transfer logic, binary numbers are stored in groups of flip flops called registers. The outputs of each register are a bundle of wires called a "bus" that carries that number to other calculations. A calculation is simply a piece of combinatorial logic. Each calculation also has an output bus, and these may be connected to the inputs of several registers.

Sometimes a register will have a multiplexer on its input, so that it can store a number from any one of several buses. Alternatively, the outputs of several items may be connected to a bus through buffers that can turn off the output of all of the devices except one. A sequential state machine controls when each register accepts new data from its input. In the 1980s, some researchers discovered that almost all synchronous register-transfer machines could be converted to asynchronous designs by using first-in-first-out synchronization logic. In this scheme, the digital machine is characterized as a set of data flows. In each step of the flow, an asynchronous "synchronization circuit" determines when the outputs of that step are valid, and presents a signal that says, "grab the data" to the stages that use that stage's inputs.

It turns out that just a few relatively simple synchronization circuits are needed. The most general-purpose register-transfer logic machine is a computer. This is basically an automatic binary abacus. The control unit of a computer is usually designed as a microprogram run by a microsequencer. A microprogram is much like a player-piano roll. Each table entry or "word" of the microprogram commands the state of every bit that controls the computer. The sequencer then counts, and the count addresses the memory or combinatorial logic machine that contains the microprogram.

The bits from the microprogram control the arithmetic logic unit, memory and other parts of the computer, including the microsequencer itself. In this way, the complex task of designing the controls of a computer is reduced to a simpler task of programming a relatively independent collection of much simpler logic machines.

Computer architecture is a specialized engineering activity that tries to arrange the registers, calculation logic, buses and other parts of the computer in the best way for some purpose. Computer architects have applied large amounts of ingenuity to computer design to reduce the cost and increase the speed and immunity to programming errors of computers. An increasingly common goal is to reduce the power used in a battery-powered computer system, such as a cell-phone. Many computer architects serve an extended apprenticeship as microprogrammers. "Specialized computers" are usually a conventional computer with a special-purpose microprogram.

AUTOMATED DESIGN TOOLS

To save costly engineering effort, much of the effort of designing large logic machines has been automated. The computer programs are called "electronic design automation tools" or just "EDA." Simple truth table-style descriptions of logic are often optimized with EDA that automatically produces reduced systems of logic gates or smaller lookup tables that still produce the desired outputs. The most common example of this kind of software is the Espresso heuristic logic minimizer.

Most practical algorithms for optimizing large logic systems use algebraic manipulations or binary decision diagrams, and there are promising experiments with genetic algorithms and annealing optimizations. To automate costly engineering processes, some EDA can take state tables that describe state machines and automatically produce a truth table or a function table for the combinatorial part of a state machine. The state table is a piece of text that lists each state, together with the conditions controlling the transitions between them and the belonging output signals.

It is common for the function tables of such computer-generated state-machines to be optimized with logic-minimization software such as Minilog. Often, real logic systems are designed as a series of sub-projects, which are combined using a "tool flow." The tool flow is usually a "script," a simplified computer language that can invoke the software design tools in the right order.

Tool flows for large logic systems such as microprocessors can be thousands of commands long, and combine the work of hundreds of engineers. Writing and debugging tool flows is an established engineering specialty in companies that produce digital designs. The tool flow usually terminates in a detailed computer file or set of files that describe how to physically construct the logic. Often it consists of instructions to draw the transistors and wires on an integrated circuit or a printed circuit board.

Parts of tool flows are "debugged" by verifying the outputs of simulated logic against expected inputs. The test tools take computer files with sets of inputs and outputs, and highlight discrepancies between the simulated behaviour and the expected behaviour. Once the input data is believed correct, the design itself must still be verified for correctness. Some tool flows verify designs by first producing a design, and then scanning the design to produce compatible input data for the tool flow. If the scanned data matches the input data, then the tool flow has probably not introduced errors.

The functional verification data are usually called "test vectors." The functional test vectors may be preserved and used in the factory to test that newly constructed logic works correctly. However, functional test patterns don't discover common fabrication faults. Production tests are often designed by software tools called "test pattern generators." These generate test vectors by examining the structure of the logic and systematically generating tests for particular faults. This way the fault coverage can closely approach 100%, provided the design is properly made testable.

Once a design exists, and is verified and testable, it often needs to be processed to be manufacturable as well. Modern integrated circuits have features smaller than the wavelength of the light used to expose the photoresist. Manufacturability software adds interference patterns to the exposure masks to eliminate open-circuits, and enhance the masks' resolution and contrast.

Design for Testability

A large logic machine (say, with more than a hundred logical variables) can have an astronomical number of possible states. Obviously, in the factory, testing every state is impractical if testing each state takes a microsecond, and there are more states than the number of microseconds since the universe began. Unfortunately, this ridiculous-sounding case is typical. Fortunately, large logic machines are almost always designed as assemblies of smaller logic machines. To save time, the smaller sub-machines are isolated by permanently-installed "design for test" circuitry, and are tested independently.

One common test scheme known as "scan design" moves test bits serially (one after another) from external test equipment through one or more serial shift registers known as "scan chains". Serial scans have only one or two wires to carry the data, and minimize the physical size and expense of the infrequently-used test logic. After all the test data bits are in place, the design is reconfigured to be in "normal mode" and one or more clock pulses are applied, to test for faults (e.g. stuck-at low or stuck-at high) and capture the test result into flip-flops and/or latches in the scan shift register(s). Finally, the result of the test is shifted out to the block boundary and compared against the predicted "good machine" result. In a board-test environment, serial to parallel testing has been formalized with a standard called "JTAG" (named after the "Joint Test Action Group" that proposed it). Another common testing scheme provides a test mode that forces some part of the logic machine to enter a "test cycle." The test cycle usually exercises large independent parts of the machine.

Trade-offs

Several numbers determine the practicality of a system of digital logic. Engineers explored numerous electronic devices to get an ideal combination of fanout, speed, low cost and reliability. The cost of a logic gate is crucial. In the 1930s, the earliest digital logic systems were constructed from telephone relays because these were inexpensive and relatively reliable. After that, engineers always used the cheapest available electronic switches that could still fulfill the requirements.

The earliest integrated circuits were a happy accident. They were constructed not to save money, but to save weight, and permit the Apollo Guidance Computer to control an inertial guidance system for a spacecraft. The first integrated circuit logic gates cost nearly $50 (in 1960 dollars, when an engineer earned $10,000/year). To everyone's surprise, by the time the circuits were mass-produced, they had become the least-expensive method of constructing digital logic. Improvements in this technology have driven all subsequent improvements in cost.

With the rise of integrated circuits, reducing the absolute number of chips used represented another way to save costs. The goal of a designer is not just to make the simplest circuit, but to keep the component count down. Sometimes this results in slightly more complicated designs with respect to the underlying digital logic but nevertheless reduces the number of components, board size, and even power consumption. For example, in some logic families, NAND gates are the simplest digital gate to build. All other logical operations can be implemented by NAND gates.

If a circuit already required a single NAND gate, and a single chip normally carried four NAND gates, then the remaining gates could be used to implement other logical operations like logical and. This could eliminate the need for a separate chip containing those different types of gates. The "reliability" of a logic gate describes its mean time between failure (MTBF). Digital machines often have millions of logic gates. Also, most digital machines are "optimized" to reduce their cost. The result is that often, the failure of a single logic gate will cause a digital machine to stop working.

Digital machines first became useful when the MTBF for a switch got above a few hundred hours. Even so, many of these machines had complex, well-rehearsed repair procedures, and would be nonfunctional for hours because a tube burned-out, or a moth got stuck in a relay. Modern transistorized integrated circuit logic gates have MTBFs of nearly a trillion (1 × 10) hours, and need them because they have so many logic gates. Fanout describes how many logic inputs can be controlled by a single logic output.

The minimum practical fanout is about five. Modern electronic logic using CMOS transistors for switches have fanouts near fifty, and can sometimes go much higher. The "switching speed" describes how many times per second an inverter (an electronic representation of a "logical not" function) can change from true to false and back. Faster logic can accomplish more operations in less time.

Digital logic first became useful when switching speeds got above fifty hertz, because that was faster than a team of humans operating mechanical calculators. Modern electronic digital logic routinely switches at five gigahertz (5 × 10 hertz), and some laboratory systems switch at more than a terahertz (1 × 10 hertz).

LOGIC FAMILIES

Design started with relays. Relay logic was relatively inexpensive and reliable, but slow. Occasionally a mechanical failure would occur. Fanouts were typically about ten, limited by the resistance of the coils and arcing on the contacts from high voltages. Later, vacuum tubes were used. These were very fast, but generated heat, and were unreliable because the filaments would burn out. Fanouts were typically five to seven, limited by the heating from the tubes' current. In the 1950s, special "computer tubes" were developed with filaments that omitted volatile elements like silicon.

These ran for hundreds of thousands of hours. The first semiconductor logic family was Resistor-transistor logic. This was a thousand times more reliable than tubes, ran cooler, and used less power, but had a very low fan-in of three. Diode-transistor logic improved the fanout up to about seven, and reduced the power. Some DTL designs used two power-supplies with alternating layers of NPN and PNP transistors to increase the fanout.

Transistor transistor logic (TTL) was a great improvement over these. In early devices, fanout improved to ten, and later variations reliably achieved twenty. TTL was also fast, with some variations achieving switching times as low as twenty nanoseconds. TTL is still used in some designs. Another contender was emitter coupled logic. This is very fast but uses a lot of power. It's now used mostly in radio-frequency circuits. Modern integrated circuits mostly use variations of CMOS, which is acceptably fast, very small and uses very little power. Fanouts of forty or more are possible, with some speed penalty.

NON-ELECTRONIC LOGIC

It is possible to construct non-electronic digital mechanisms. In principle, any technology capable of representing discrete states and representing logic operations could be used to build mechanical logic. Danny Hillis, co-author of The Connection Machine, once built a working computer from Tinker toys, string, a brick, and a sharpened pencil, which is supposed to be in the Houston Museum of Natural Science. Hydraulic, pneumatic and mechanical versions of logic gates exist and are used in situations where electricity cannot be used.

The first two types are considered under the heading of fluidics. One application of fluidic logic is in military hardware that is likely to be exposed to a nuclear electromagnetic pulse (nuclear EMP, or NEMP) that would destroy electrical circuits. Mechanical logic is frequently used in inexpensive controllers, such as those in washing machines. Famously, the first computer design, by Charles Babbage, was designed to use mechanical logic. Mechanical logic might also be used in very small computers that could be built by nanotechnology. Another example is that if two particular enzymes are required to prevent the construction of a particular protein, this is the equivalent of a biological "NAND" gate.

Recent Developments

The discovery of superconductivity has enabled the development of Rapid Single Flux Quantum (RSFQ) circuit technology, which uses Josephson junctions instead of transistors. Most recently, attempts are being made to construct purely optical computing systems capable of processing digital information using nonlinear optical elements.

CIRCUIT DIAGRAM

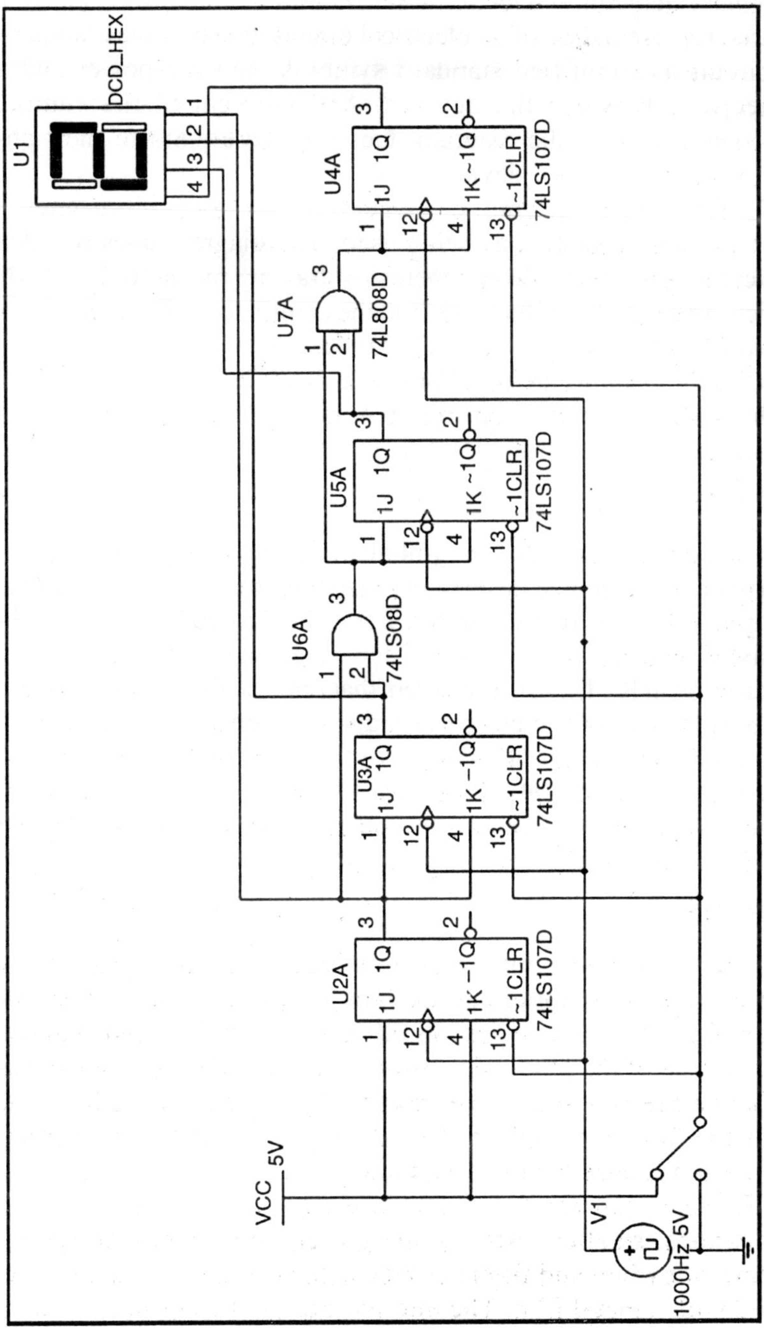

Fig. The Circuit Diagram for a 4 Bit TTL Counter, a Type of State Machine

A circuit diagram (also known as an electrical diagram, wiring diagram, elementary diagram, or electronic schematic) is a simplified conventional pictorial representation of an electrical circuit. It shows the components of the circuit as simplified standard symbols, and the power and signal connections between the devices. Arrangement of the components interconnections on the diagram does not correspond to their physical locations in the finished device.

Unlike a block diagram or layout diagram, a circuit diagram shows the actual wire connections being used. The diagram does not show the physical arrangement of components. A drawing meant to depict what the physical arrangement of the wires and the components they connect is called "artwork" or "layout" or the "physical design."

Circuit diagrams are used for the design (circuit design), construction (such as PCB layout), and maintenance of electrical and electronic equipment.

LEGENDS

On a circuit diagram, the symbols for components are labelled with a descriptor (or reference designator) matching that on the list of parts. For example, C1 is the first capacitor, L1 is the first inductor, Q1 is the first transistor, and R1 is the first resistor (note that this is not written as a subscript, as in R_1, L_1,...). The letters that precede the numbers were chosen in the early days of the electrical industry, even before the vacuum tube (thermionic valve), so "Q" was the only one available for semiconductor devices in the mid-twentieth century. Often the value or type designation of the component is given on the diagram beside the part, but detailed specifications would go on the parts list.

SYMBOLS

Circuit diagram symbols have differed from country to country and have changed over time, but are now to a large extent internationally standardized. Simple components often had symbols intended to represent some feature of the physical construction of the device. For example, the symbol for a resistor shown here dates back to the days when that component was made from a long piece of wire wrapped in such a manner as to not produce inductance, which would have made it a coil.

These wirewound resistors are now used only in high-power applications, smaller resistors being cast from *carbon composition* (a mixture of carbon and filler) or fabricated as an insulating tube or chip coated with a metal film. The internationally standardized symbol for a resistor is therefore now simplified to an oblong, sometimes with the value in ohms written inside, instead of the zig-zag symbol. A less common symbol is simply a series of peaks on one side of the line representing the conductor, rather than back-and-forth as shown here.

STANDARDS

There are several national and international standards for graphical symbols in circuit diagrams, in particular:

- IEC 60617 (also known as British Standard BS 3939)
- ANSI standard Y32 (also known as IEEE Std 315)

IEC 60617 originally consisted of 13 parts, from resistors and capacitors to logic symbols and even a generalised drawing standard of connections and bus line widths. It is now published as a subscription online database IEC 60617-DB.

Different symbols may be used depending on the discipline using the drawing; for example, lighting and power symbols used as part of architectural drawings may be different from symbols for devices used in electronics.

LINKAGES

Schematic wire junctions:

- Old style: (a) connection, (b) no connection.
- One CAD style: (a) connection, (b) no connection.
- Alternative CAD Style: (a) connection, (b) no connection.

The linkages between leads were once simple crossings of lines; one wire insulated from and "jumping over" another was indicated by it making a little semicircle over the other line. With the arrival of computerized drafting, a connection of two intersecting wires was shown by a crossing with a dot or "blob", and a crossover of insulated wires by a simple crossing without a dot.

However, there was a danger of confusing these two representations if the dot was drawn too small or omitted. Modern practice is to avoid using the "crossover with dot" symbol, and to draw the wires meeting at two points instead of one. It is also common to use a hybrid style, showing connections as a cross with a dot while insulated crossings use the semicircle.

The following codes which vary slightly from the American codes are in common use in European and Australian standard electrical circuit diagrams. These codes are used for the "reference designators" printed on PCBs (which match the corresponding ones written on the corresponding schematic).

- A: Assemblies
- B: Transducers (photo cells, inductive proximity, thermocouple, flame detection)

- C: Capacitors
- D: Storage devices
- E: Miscellaneous
- F: Fuses
- G: Generator, battery pack
- H: Indicators, lamps (not for illumination), signalling devices
- K: Relays, contactors
- L: Inductors and filters
- M: Motors
- N: Analogue devices
- P: Measuring/test equipment
- Q: Circuit breakers, isolators, re-closers
- R: Resistors, brake resistors
- S: Switches, push buttons, emergency stops and limit switches
- T: Transformers
- U: Power converters, variable speed drives, soft starters, DC power supplies
- V: Semiconductors
- W: Wires, conductors, power, neutral and earthing busses
- X: Terminal strips, terminations, joins
- Y: Solenoids, electrical actuators
- Z: Filters

Detailed rules for reference designations are provided in the International standard IEC 61346.

ORGANIZATION OF DRAWINGS

It is a usual although not universal convention that schematic drawings are organized on the page from left to right and top to bottom in the same sequence as the flow of the main signal or power path. For example, a schematic for a radio receiver might start with the antenna input at the left of the page and end with the loudspeaker at the right.

Positive power supply connections for each stage would be shown towards the top of the page, with grounds, negative supplies, or other return paths towards the bottom. Schematic drawings intended for maintenance may have the principle signal paths highlighted to assist in understanding the signal flow through the circuit. More complex devices have multi-page schematics and must rely on cross-reference symbols to show the flow of signals between the different sheets of the drawing.

Detailed rules for the preparation of circuit diagrams (and other document kinds used in electrotechnology) are provided in the International standard IEC 61082-1. Relay logic line diagrams (also called ladder logic diagrams) use another common standardized convention for organizing schematic drawings, with a vertical power supply "rail" on the left and another on the right, and components strung between them like the rungs of a ladder.

ART WORK

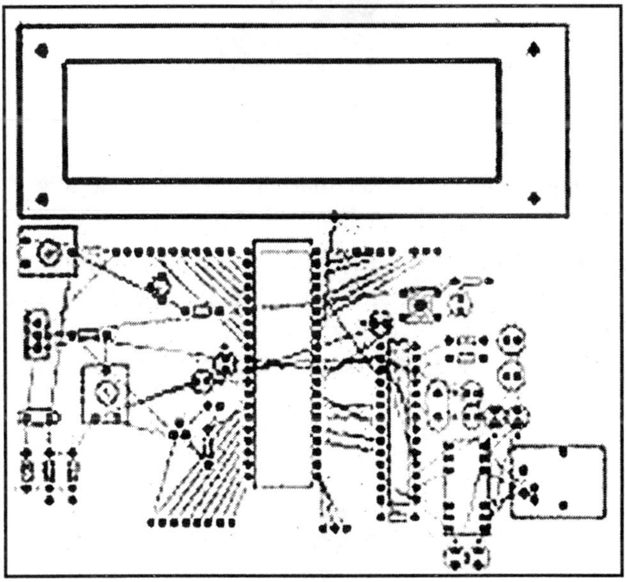

Fig. A Rat's Nest

Once the schematic has been made, it is converted into a layout that can be fabricated onto a Printed Circuit Board (PCB). The layout is usually prepared by the process of schematic capture. The result is what is known as a Rat's Nest. The Rat's Nest is a jumble of wires (lines) criss crossing each other to their destination nodes. These wires are routed either manually or by the use of Electronics Design Automation (EDA) tools. The EDA tools arrange and rearrange the placement of components and finds paths for tracks to connect various nodes. This results into an Art Work.

A generalized design flow would be as:

Schematic → Schematic Capture → Rat's Nest → Routing → Art Work → PCB Development & etching → Component Mounting → Testing

Diode Bridge

A diode bridge or bridge rectifier is an arrangement of four diodes in a bridge configuration that provides the same polarity of output voltage for either polarity of input voltage. When used in its most common application, for conversion of alternating current (AC) input into direct current (DC) output, it is known as a bridge rectifier.

Fig. Three Bridge Rectifiers. The Size is Generally Related to the Current Handling Capability.

A bridge rectifier provides full-wave rectification from a two-wire AC input, resulting in lower cost and weight as compared to a centre-tapped transformer design, but has two diode drops rather than one, thus exhibiting reduced efficiency over a centre-tapped design for the same output voltage.

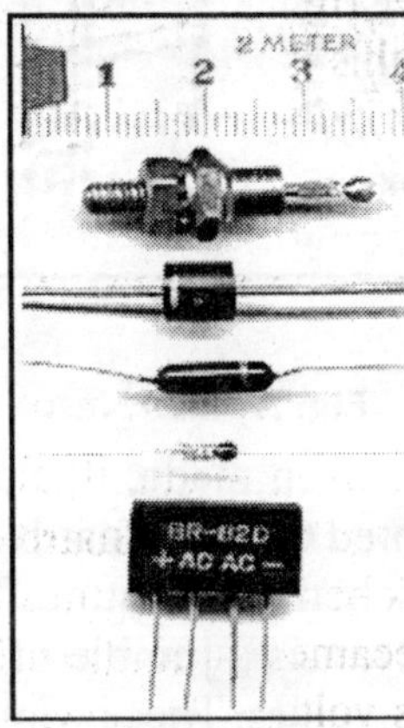

Fig. Diodes; the One on the Bottom is a Diode Bridge

The essential feature of a diode bridge is that the polarity of the output is the same regardless of the polarity at the input. The diode bridge circuit is also known as the *Graetz circuit* after its inventor, physicist Leo Graetz.

BASIC OPERATION

When the input connected at the left corner of the diamond is positive with respect to the one connected at the right hand corner, current flows to the right along the upper coloured path to the output, and returns to the input supply via the lower one. When the right hand corner is positive relative to the left hand corner, current flows along the upper coloured path and returns to the supply via the lower coloured path.

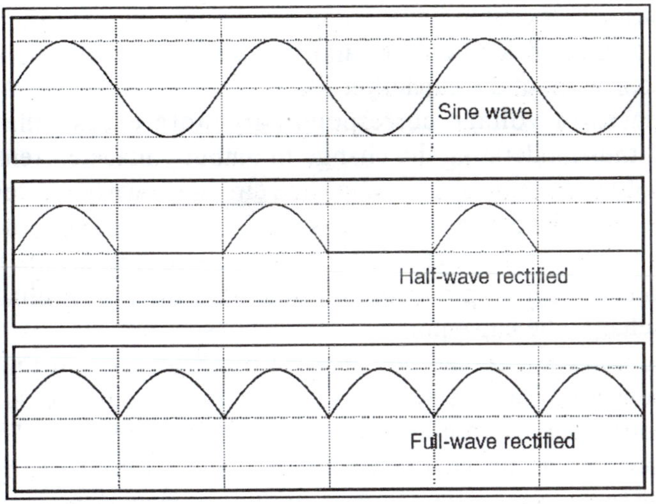

Fig. AC, Half-wave and Full Wave Rectified Signals

In each case, the upper right output remains positive with respect to the lower right one. Since this is true whether the input is AC or DC, this circuit not only produces DC power when supplied with AC power: it also can provide what is sometimes called "reverse polarity protection". That is, it permits normal functioning when batteries are installed backwards or DC input-power supply wiring "has its wires crossed" (and protects the circuitry it powers against damage that might occur without this circuit in place).

Prior to availability of integrated electronics, such a bridge rectifier was always constructed from discrete components. Since about 1950, a single four-terminal component containing the four diodes connected in the bridge configuration became a standard commercial component and is now available with various voltage and current ratings.

OUTPUT SMOOTHING

For many applications, especially with single phase AC where the full-wave bridge serves to convert an AC input into a DC output, the addition of a capacitor may be important because the bridge alone supplies an output voltage of fixed polarity but pulsating magnitude.

The function of this capacitor, known as a reservoir capacitor (aka smoothing capacitor) is to lessen the variation in (or 'smooth') the rectified AC output voltage waveform from the bridge. One explanation of 'smoothing' is that the capacitor provides a low impedance path to the AC component of the output, reducing the AC voltage across, and AC current through, the resistive load. In less technical terms, any drop in the output voltage and current of the bridge tends to be cancelled by loss of charge in the capacitor. This charge flows out as additional current through the load. Thus the change of load current and voltage is reduced relative to what would occur without the capacitor.

Increases of voltage correspondingly store excess charge in the capacitor, thus moderating the change in output voltage/current. Also see rectifier output smoothing. The simplified circuit shown has a well deserved reputation for being dangerous, because, in some applications, the capacitor can retain a *lethal* charge after the AC power source is removed. If supplying a dangerous voltage, a practical circuit should include a reliable way to safely discharge the capacitor.

If the normal load can not be guaranteed to perform this function, perhaps because it can be disconnected, the circuit should include a bleeder resistor connected as close as practical across the capacitor. This resistor should consume a current large enough to discharge the capacitor in a reasonable time, but small enough to minimise unnecessary power waste. Because a bleeder sets a minimum current drain, the regulation of the circuit, defined as percentage voltage change from minimum to maximum load, is improved. However in many cases the improvement is of insignificant magnitude.

The capacitor and the load resistance have a typical time constant $\tau = RC$ where C and R are the capacitance and load resistance respectively. As long as the load resistor is large enough so that this time constant is much longer than the time of one ripple cycle, the above configuration will produce a smoothed DC voltage across the load. In some designs, a series resistor at the load side of the capacitor is added. The smoothing can then be improved by adding additional stages of capacitor–resistor pairs, often done only for sub-supplies to critical high-gain circuits that tend to be sensitive to supply voltage noise.

The idealized waveforms shown above are seen for both voltage and current when the load on the bridge is resistive. When the load includes a smoothing capacitor, both the voltage and the current waveforms will be greatly changed. While the voltage is smoothed, as described above, current will flow through the bridge only during the time when the input voltage is greater than the capacitor voltage. For example, if the load draws an average current of n Amps, and the diodes conduct for 10% of the time, the average diode current during conduction must be 10n Amps. This non-sinusoidal current leads to harmonic distortion and a poor power factor in the AC supply.

In a practical circuit, when a capacitor is directly connected to the output of a bridge, the bridge diodes must be sized to withstand the current surge that occurs when the power is turned on at the peak of the AC voltage and the capacitor is fully discharged. Sometimes a small series resistor is included before the capacitor to limit this current, though in most applications the power supply transformer's resistance is already sufficient. Output can also be smoothed using a choke and second capacitor. The choke tends to keep the current (rather than the voltage) more constant. Due to the relatively high cost of an effective choke compared to a resistor and capacitor this is not employed in modern equipment.

Some early console radios created the speaker's constant field with the current from the high voltage ("B +") power supply, which was then routed to the consuming circuits, (permanent magnets were then too weak for good performance) to create the speaker's constant magnetic field. The speaker field coil thus performed 2 jobs in one: it acted as a choke, filtering the power supply, and it produced the magnetic field to operate the speaker.

POLYPHASE DIODE BRIDGES

This construction can be generalized to rectify Polyphase system AC inputs. For instance, for *three*-phase AC, a full wave bridge rectifier consists of six diodes.

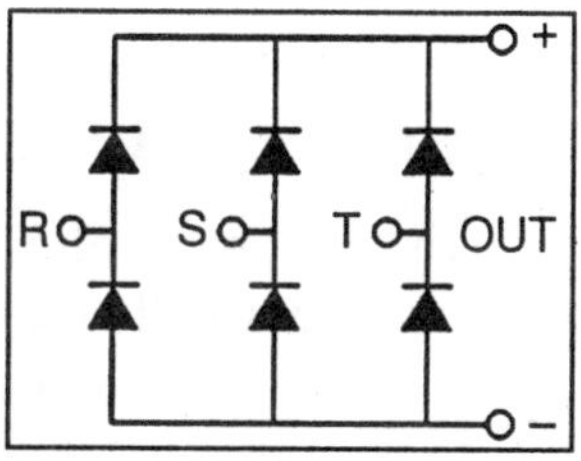

Fig. Three Phase Bridge Rectifier for a Wind Turbine.

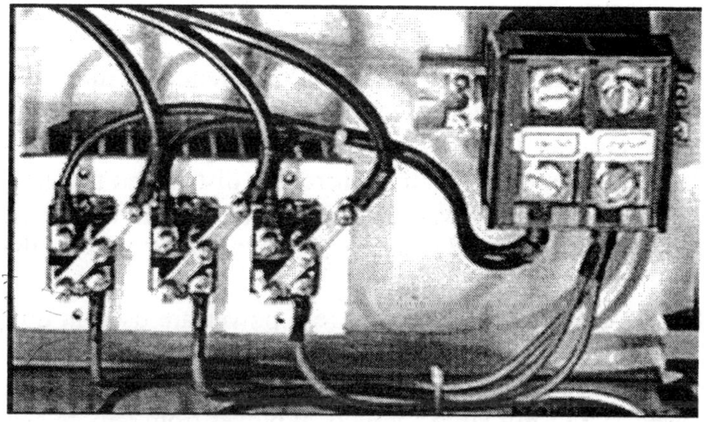

Fig. Three Phase Bridge Rectifier for a Wind Turbine.

Chapter 7

Direct Current

Direct current (DC) is the unidirectional flow of electric charge. Direct current is produced by such sources as batteries, thermocouples, solar cells, and commutator-type electric machines of the dynamo type. Direct current may flow in a conductor such as a wire, but can also be through semiconductors, insulators, or even through a vacuum as in electron or ion beams. In direct current, the electric charges flow in a constant direction, distinguishing it from alternating current (AC). A term formerly used for *direct current* was Galvanic current.

TYPES OF DIRECT CURRENT

Direct current may be obtained from an alternating current supply by use of a current-switching arrangement called a rectifier, which contains electronic elements (usually) or electromechanical elements (historically) that allow current to flow only in one direction. Direct current may be made into alternating current with an inverter or a motor-generator set. The first commercial electric power transmission (developed by Thomas Edison in the late nineteenth century) used direct current.

Because of the advantage of alternating current over direct current in transforming and transmission, electric power distribution today is nearly all alternating current. For applications requiring direct current, such as third rail power systems, alternating current is distributed to a substation, which utilizes a rectifier to convert the power to direct current. Direct current is used to charge batteries, and in nearly all electronic systems as the power supply.

Very large quantities of direct-current power are used in production of aluminum and other electrochemical processes. Direct current is used for some railway propulsion, especially in urban areas. High voltage direct current is used to transmit large amounts of power from remote generation sites or to interconnect alternating current power grids.

VARIOUS DEFINITIONS

Within electrical engineering, the term *DC* is used to refer to power systems that use only one polarity of voltage or current, and to refer to the constant, zero-frequency, or slowly varying local mean value of a voltage or current. For example, the voltage across a DC voltage source is constant as is the current through a DC current source. The DC solution of an electric circuit is the solution where all voltages and currents are constant. It can be shown that any stationary voltage or current waveform can be decomposed into a sum of a DC component and a zero-mean time-varying component; the DC component is defined to be the expected value, or the average value of the voltage or current over all time.

Although DC stands for "Direct *Current*", DC sometimes refers to "constant polarity." With this definition, DC voltages can vary in time, such as the raw output of a rectifier or the fluctuating voice signal on a telephone line. Some forms of DC (such as that produced by a voltage regulator) have almost no variations in voltage, but may still have variations in output power and current.

APPLICATIONS

Direct-current installations usually have different types of sockets, switches, and fixtures, mostly due to the low voltages used, from those suitable for alternating current. It is usually important with a direct-current appliance not to reverse polarity unless the device has a diode bridge to correct for this (most battery-powered devices do not). DC is commonly found in many low-voltage applications, especially where these are powered by batteries, which can produce only DC, or solar power systems, since solar cells can produce only DC.

Most automotive applications use DC, although the alternator is an AC device which uses a rectifier to produce DC. Most electronic circuits require a DC power supply. Applications using fuel cells (mixing hydrogen and oxygen together with a catalyst to produce electricity and water as byproducts) also produce only DC. Many telephones connect to a twisted pair of wires, and internally separate the AC component of the voltage between the two wires (the audio signal) from the DC component of the voltage between the two wires (used to power the phone).

Telephone exchange communication equipment, such as DSLAM, uses standard –48V DC power supply. The negative polarity is achieved by grounding the positive terminal of power supply system and the battery bank. This is done to prevent electrolysis depositions. An electrified third rail can be used to power both underground (subway) and overground trains.

BIASING (ELECTRONICS)

Biasing in electronics is the method of establishing predetermined voltages and/or currents at various points of a circuit to set an appropriate *operating point*. The operating point of a device, also known as *bias point* or *quiescent point* (or simply *Q-point*), is the DC voltage and/or current which, when applied to a device, causes it to operate in a certain desired fashion. The term is normally used in connection with devices such as transistors and diodes which are used in amplification or rectification.

IMPORTANCE IN LINEAR CIRCUITS

Linear circuits involving transistors typically require specific DC voltages and currents to operate correctly, which can be achieved using a biasing circuit. As an example of the need for careful biasing, consider a transistor amplifier. In linear amplifiers, a small input signal gives larger output signal without any change in shape (low distortion): the input signal causes the output signal to vary up and down about the Q-point in a manner strictly proportional to the input. However, because a transistor is nonlinear, the transistor amplifier only approximates linear operation.

For low distortion, the transistor must be biased so the output signal swing does not drive the transistor into a region of extremely nonlinear operation. For a bipolar transistor amplifier, this requirement means that the transistor must stay in the *active* mode, and avoid cut-off or saturation. The same requirement applies to a MOSFET amplifier, although the terminology differs a little: the MOSFET must stay in the *active* mode (or saturation mode), and avoid cut-off or Ohmic operation (or triode mode). More detail follows.

BIPOLAR JUNCTION TRANSISTORS

For bipolar junction transistors the bias point is chosen to keep the transistor operating in the active mode, using a variety of circuit techniques, establishing the Q-point DC voltage and current. A small signal is then applied on top of the Q-point bias voltage, thereby either modulating or switching the current, depending on the purpose of the circuit. The quiescent point of operation is typically near the middle of DC *load line*. The process of obtaining certain DC collector current at a certain DC collector voltage by setting up operating point is called biasing.

After establishing the operating point, when input signal is applied, the output signal should not move the transistor either to saturation or to cut-off. However, this unwanted shift still might occur, due to the following reasons:

- Parameters of transistors depend on junction temperature. As junction temperature increases, leakage current due to minority charge carriers (I_{CBO}) increases. As I_{CBO} increases, I_{CEO} also increases, causing an increase in collector current I_C. This produces heat at the collector junction. This process repeats, and, finally, Q-point may shift into the saturation region. Sometimes, the excess heat produced at the junction may even burn the transistor. This is known as thermal runaway.

- When a transistor is replaced by another of the same type, the Q-point may shift, due to changes in parameters of the transistor, such as *current gain* (β) which varies slightly for each unique transistor.

To avoid a shift of Q-point, bias-stabilization is necessary. Various biasing circuits can be used for this purpose.

NETWORK ANALYSIS (ELECTRICAL CIRCUITS)

A network, in the context of electronics, is a collection of interconnected components. Network analysis is the process of finding the voltages across, and the currents through, every component in the network. There are a number of different techniques for achieving this. However, for the most part, they assume that the components of the network are all linear. The methods described in this article are only applicable to *linear* network analysis except where explicitly stated.

Component: A device with two or more terminals into which, or out of which, charge may flow.

Node: A point at which terminals of more than two components are joined. A conductor with a substantially zero resistance is considered to be a node for the purpose of analysis.

Branch: The component(s) joining two nodes.

Mesh: A group of branches within a network joined so as to form a complete loop.

Port: Two terminals where the current into one is identical to the current out of the other.

Circuit: A current from one terminal of a generator, through load component(s) and back into the other terminal. A circuit is, in this sense, a one-port network and is a trivial case to analyse. If there is any connection to any other circuits then a non-trivial network has been formed and at least two ports must exist.

Transfer function: The relationship of the currents and/or voltages between two ports. Most often, an input port and an output port are discussed and the transfer function is described as gain or attenuation.

Component transfer function: For a two-terminal component (i.e. one-port component), the current and voltage are taken as the input and output and the transfer function will have units of impedance or admittance (it is usually a matter of arbitrary convenience whether voltage or current is considered the input). A three (or more) terminal component effectively has two (or more) ports and the transfer function cannot be expressed as a single impedance. The usual approach is to express the transfer function as a matrix of parameters. These parameters can be impedances, but there is a large number of other approaches, see two-port network.

EQUIVALENT CIRCUITS

A useful procedure in network analysis is to simplify the network by reducing the number of components. This can be done by replacing the actual components with other notional components that have the same effect. A particular technique might directly reduce the number of components, for instance by combining impedances in series.

On the other hand it might merely change the form in to one in which the components can be reduced in a later operation. For instance, one might transform a voltage generator into a current generator using Norton's theorem in order to be able to later combine the internal resistance of the generator with a parallel impedance load.

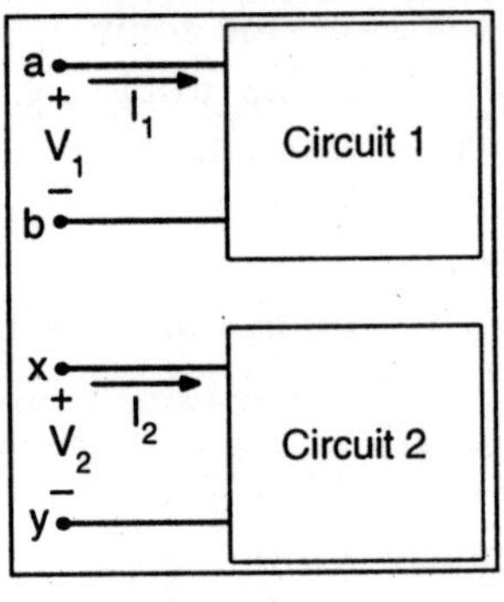

Fig. Circuits

A resistive circuit is a circuit containing only resistors, ideal current sources, and ideal voltage sources.

If the sources are constant (DC) sources, the result is a DC circuit. The analysis of a circuit refers to the process of solving for the voltages and currents present in the circuit. The solution principles outlined here also apply to phasor analysis of AC circuits.

Two circuits are said to be equivalent with respect to a pair of terminals if the voltage across the terminals and current through the terminals for one network have the same relationship as the voltage and current at the terminals of the other network.

If $V_2 = V_1$ implies $I_2 = I_1$ for all (real) values of V_1, then with respect to terminals ab and xy, circuit 1 and circuit 2 are equivalent.

The above is a sufficient definition for a one-port network. For more than one port, then it must be defined that the currents and voltages between all pairs of corresponding ports must bear the same relationship. For instance, star and delta networks are effectively three port networks and hence require three simultaneous equations to fully specify their equivalence.

IMPEDANCES IN SERIES AND IN PARALLEL

Any two terminal network of impedances can eventually be reduced to a single impedance by successive applications of impendances in series or impendances in parallel.

DELTA-WYE TRANSFORMATION

A network of impedances with more than two terminals cannot be reduced to a single impedance equivalent circuit. An n-terminal network can, at best, be reduced to n impedances.

For a three terminal network, the three impedances can be expressed as a three node delta (Δ) network or a four node star (Y) network. These two networks are equivalent and the transformations between them are given below. A general network with an arbitrary number of terminals cannot be reduced to the minimum number of impedances using only series and parallel combinations.

In general, Y–Δ and Δ–Y transformations must also be used. It can be shown that this is sufficient to find the minimal network for any arbitrary network with successive applications of series, parallel, Y–Δ and Δ–Y; no more complex transformations are required. For equivalence, the impedances between any pair of terminals must be the same for both networks, resulting in a set of three simultaneous equations. The equations below are expressed as resistances but apply equally to the general case with impedances.

SOURCE TRANSFORMATION

A generator with an internal impedance (ie non-ideal generator) can be represented as either an ideal voltage generator or an ideal current generator plus the impedance.

These two forms are equivalent and the transformations are given below. If the two networks are equivalent with respect to terminals ab, then V and I must be identical for both networks.

SWITCHING AND SWITCHES

MANUAL SWITCHES

The function of this switch is evident from its appearance. It is used to make and break electric connections. Such a switch might be used in the simple circuit to turn on or off the light bulb.

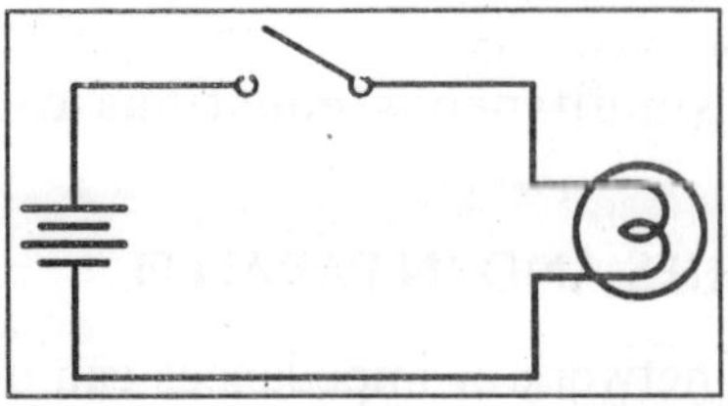

Fig. Circuit to Turn on and off a Light Bulb.

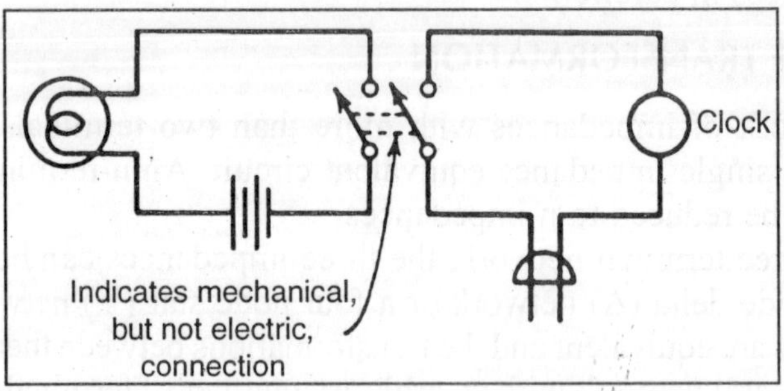

Fig. Circuit to Turn on and off two Separate Subcircuits Simultaneously

Another common switch that performs the same function is the toggle switch. In this type there are two terminals extending from the bottom, and the switch either connects them together or disconnects them, depending upon which way the handle is thrown. Neither the body of the switch nor the handle is electrically connected to either of the two terminals. Any switch that makes or breaks one pair of contacts is called a single-pole, single-throw (or SPST) switch.

Frequently more than one circuit must be turned on or off at the same time. For example, consider the circuit. This circuit is intended to measure the total time the light bulb is on. Therefore, the clock and the bulb should be connected and disconnected simultaneously. It consists essentially of two knife switches connected together mechanically, but not electrically. The same handle operates both switches. Note that there are four terminals extending from the bottom of this toggle switch. The terminals on toggle switches are arranged in the same way as for the corresponding knife switch. Thus, on this switch, the actual terminals would be connected. Any switch that makes or breaks two pairs of contacts is called a *double-pole*, single-throw (or DPST) switch.

The simple "on-off" switch first mentioned in this chapter may be said to perform a "yes-no" operation. The bulb is either lit or not lit. The second—double—switch performs an "and" operation as well; that is, the switch causes bulb *A* and *B* to light or not to light.

Another common form of switching requirement is the "or" operation, in which either one circuit *or* another is activated. A good example of the "or" circuit is the one used for probability-matching experiments, in which one or the other of two lights will be turned on, and the subject is required to guess which it will be. Any switch that connects a single terminal with either one or the other of two terminals is called a single-pole, double-throw (or SPDT) switch. The first switch mentioned in this chapter was designated as a single pole, single-throw (or SPST) switch.

It is single-pole because it operates on only one circuit, and it is called single-throw because that circuit may be either on or off. (This is clearer when contrasted with other types of switches.) The second type of switch described was the double-pole, single-throw (or DPST) switch, so called because two electrically independent circuits are either on or off. The "or" switch is called single-pole, double-throw (SPDT). If instead of selecting one or the other of two bulbs, one of four were to be chosen, an SP4T switch would be required. (Note that this switch has a mechanical structure different from the switches discussed thus far. The common contact rotates across the others, and the switch is called a rotary switch.)

The microswitches on your board are SPDT switches. Note that there are two important differences between the knife and the microswitch. First, the knife switch has no spring in it, so it stays where it is put, whereas the microswitch snaps back to its resting position as soon as pressure is released. The microswitch is therefore said to give "momentary" contact. Second, whereas the knife switch may be left in a middle position where neither circuit is closed, the microswitch is always in either one position or the other (except for an extremely short transit time). Because of these two factors, one of the terminals on the switch is labeled "normally closed" (or NC) and another "normally open" (or NO).

This means that the unlabeled, or "common," terminal is *normally* connected to the NC terminal, but when the handle is pressed, the common terminal shifts over and connects with the no terminal. When the pressure is released, it shifts back to the NC terminal. The toggle switch on your board is a double-pole, double-throw (or DPDT) switch. This versatile contact configuration may be used as a SPST, SPDT, DPST or DPDT switch simply by using the appropriate terminals. The DPDT switch might be used as such when either light *A* or light *B* must be on and the "on" time for each must be recorded.

In the preceding discussion and figures, knife switches are used to illustrate the various classes of switching action because their functions are clear from their structure. However, knife switches are rarely used when actual circuits are constructed. They are bigger, less positive in action, and more exposed than toggle switches. Almost the only time they are required in behavioural instrumentation is when large amounts of current are to be turned on and off (large knife switches can carry larger currents than most toggle switches).

Single switches of the types discussed above may be used when *either* one *or* another *element* is to operate (e.g., SPDT), or when one *element and* another operate (e.g., DPST), or combinations of "either or" and "and" circuits. A different class of switching operations requires the use of combinations of separate switches. For example, suppose that a bulb is to be lighted when either one or another of two switches is closed. Whenever a *single circuit element or group of elements* is to operate when *either* one *or* another of two *switches* is thrown, the two switches may be connected in parallel.

This frequently occurs in apparatus used in situations where the subject normally controls the events but the experimenter wants to be able to control them as well. For example, an animal in a Skinner box may get reinforced for pressing a bar, but the experimenter would like to deliver a reinforcement when he chooses as well. In almost all Skinner boxes, the bar is mechanically connected in such a way that pressure on it operates a microswitch. Therefore, in order to permit the experimenter to control the same events that the rat controls, a switch similar to the one which is closed by a bar press is connected in parallel with the bar switch and mounted on the experimenter's control panel.

Suppose, instead, that the bar press opens a normally closed switch, and that the experimenter wishes to be able to perform the same operation manually. In this case, the experimenter may use a second, normally closed, switch connected in *series* with the animal's switch. Whenever each and any one of two or more switches is to perform the same operation, the switches are all connected in parallel if they are normally open, and all connected in series if they are normally closed.

Now consider the case where an element or set of elements is only to operate if *all* of two or more switches are closed at the same time. For example, in the Skinner box it is usually necessary for the experimenter to be able to determine when a bar press will be reinforced. In this case, the two switches are normally open and connected in series. If, instead, the switches are normally closed and a bulb is to go off only when all are open at the same time, the switches must be in parallel.

Design and build the following circuits, soldering all connections. (Of course, you do not need to solder the connections to the clips on the battery.) When performing these exercises, *always draw a complete circuit diagram on paper before you make any electric connections.* Even if this seems unnecessary to you, it constitutes practice that will be indispensable in designing and building more complicated circuits.

PROBLEM. Either bulb (6 volts) or motor (3 volts) on. (To run the motor, connect 3 volts between the two leads or terminals mounted on the motor. Most toy motors are supplied with leads. If these leads are bent frequently, they will break off at the motor end. To avoid this, connect each lead to a separate terminal on one of the mounted terminal strips. Thereafter, any connection to the motor may be made by soldering wires to the terminal strip.)

Solution. Virtually every problem in electric circuit design has a number of correct solutions.

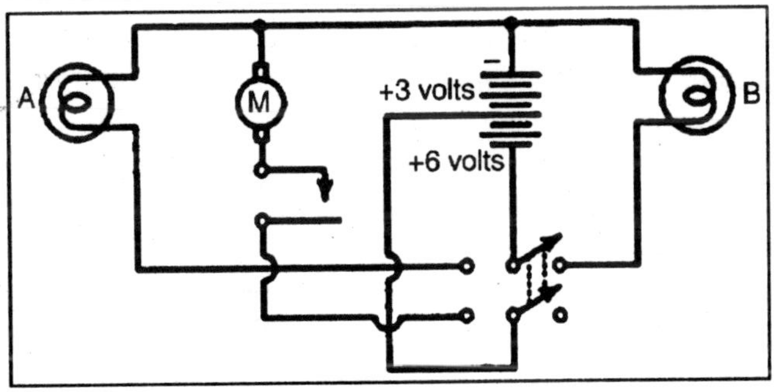

Fig. A Solution to the Problem of Either Bulb *A* or Bulb *B* on.

The descriptions and designations given above apply to all kinds of switches. That is, a 3PDT switch may be a knife switch, a toggle switch, etc. These designations merely state what might be called the logical operations of the switch. The differences between types of switches within any one *operational category* (e.g., DPDT) are related to the amounts of current and voltage that they may safely handle, their size, etc. There are so many switch shapes that it is not worthwhile to begin to enumerate them.

If you cannot actually see the insides of a switch to ascertain what sorts of connections it makes, it is usually possible to guess quite accurately from looking at the terminals of the switch. For example, if there are six terminals extending from a strange switch, it is a very good bet that it is a DPDT switch. To be sure, it is always possible to check just which terminals are connected together by putting an ohmmeter between the various pairs of terminals and operating the handle. If there is an infinite resistance between two terminals when the switch is in one position, and a very low resistance when the switch is changed, the change connected the two terminals.

SOLENOIDS

It has already been pointed out, in the discussion of meters, that current flowing through a coil of wire generates a magnetic field around the coil. A useful form of such an electromagnet is called a solenoid. This is a coil of wire wrapped around a metal core which has a hole in it.

A rod of iron slides in and out of the hole. If the rod is initially just part way in the hole and current is passed through the coil, the rod will be pulled into the core. Thus, if the handle of a camera shutter were attached to the rod for example, the shutter could be operated by remote control.

Relays

A solenoid permits the performance of a mechanical operation by remote control. If the iron rod of a solenoid were mechanically connected to the handle of a toggle switch, the switch could then be operated by remote control. This is the principle of operation of a class of extraordinarily useful devices called relays. In general, a relay is a switch that is operated electrically instead of by hand.

The part labeled "coil" consists of a large number of turns of wire wrapped around an iron core. There are two, and only two, wires or terminals coming from this coil. They are connected one to each end of the wire forming the coil. When a current is passed through the coil (by connecting a voltage between the two terminals), the iron core becomes an electromagnet just as the coil in a meter movement forms a magnet. The magnetic field produced pulls on the armature, or clapper, and the armature acts like the handle of a switch, changing the connections between the contacts. The standard relay behaves more like a microswitch, or "momentary" switch, than like a standard toggle switch, in that the contacts normally stay in one position, and are in the other position only while the armature is actually being held by the magnetic field.

As soon as current stops flowing through the coil, the spring returns the armature and the contacts to their resting or "normal" position. The relay shown in this figure is a single-pole, double-throw relay. In other words, it performs the same operations as a SPDT switch. Note the similarity between the three representations. Almost all relays consist of a magnet coil and a set of contacts, some of which are moved by the magnetic field. These two parts are *not* electrically connected together when the relay comes from the factory. The only connections between the contacts and the coil are mechanical and magnetic ones.

For some particular circuit requirements it is necessary to make an electric connection between the coil and some of the contacts, but such a connection must be made by you. It is not there to start with. This statement is emphasized because it is an extremely common error for people learning circuit design to assume that such a connection is built into the relay. Relays may be purchased with coils suitable for all sorts of currents and voltages, and with contact arrangements of an almost infinite variety. Circuits consisting of virtually nothing but relays can be made to do an enormous number of things.

For example, many of the early, large digital computers used relay circuits exclusively to perform their mathematical operations. The vacuum tubes, transistors, ferrite cores, etc., which are now being used in computers function like very fast and small relays. The rest of this chapter will consider some of the ways in which relays are used. Relays are sometimes employed to permit a very weak signal to control a much stronger signal. For example, suppose that a light is to be turned on in a room whenever it gets dark outside. Photocells will be discussed in more detail, but for this purpose it is sufficient to state that certain types of cells will generate a voltage which is proportional to the light intensity incident on them; that is, they act like variable-voltage batteries.

However, the power output of photocells is relatively small. A photocell would have to be enormous and expensive in order to deliver enough power to light a light bulb; but many photocells will deliver enough power to operate a sensitive relay. If the photocell is connected across the coil of a sensitive relay, the photocell will drive current through the relay. When the incident light exceeds some critical level, sufficient current will be driven through the relay to pull in its armature. The contacts on the relay, in turn, may handle enough power to light the bulb. When the circuit is connected, a strong light on the photocell will make the lamp turn on. In this manner, the small power of the photocell controls the relatively large power of the lamp.

If the photocell is placed outside the room, and the bulb inside, then, the bulb will turn on whenever it is light outside. But the original problem was to turn on the bulb when it got *dark* outside. In the first case, the bulb was connected across a normally open pair of contacts on the relay. Now it is connected across a normally closed pair, so that the bulb goes on when the relay goes off, and vice versa. In this circuit diagram, a DPDT relay is shown, although only one set of contacts is used. Actually, all that is needed for this circuit is a normally closed SPDT relay.

The DPDT relay was deliberately shown to point out what is perhaps obvious—that for any given circuit, the particular contact configuration required specifies the *minimum* contact requirement for the relay; any relay with *at least* this number of contacts will be satisfactory. In fact, it is usually a good idea to have extra contacts on a relay because, should it become necessary to add elements to a circuit which were not originally anticipated, these elements may often be connected through the extra contacts.

A second common use of relays is to convert a single-pole switch into a multiple-pole switch. Consider a typical rat T-maze apparatus. Suppose that, as soon as the rat leaves the start box, a clock is to start and a light is to go on. The floor is hinged, and one end rests on the arm of a microswitch.

The sensitivity of the microswitch and the lever arrangement are such that the weight of the rat on the floor will close the microswitch.

In the T-maze, this kind of arrangement can be built into the start-box floor. Then, as soon as the rat leaves the start box, the *normally closed* contacts *close*. If the microswitch were a DPDT switch, then two such circuits would be closed as soon as the rat left the start box; one circuit might be used to start the clock and the other to turn on the light. However, microswitches more complex than SPDT are expensive and generally hard to get. It is easier and cheaper to use the normally closed contacts of an ordinary SPDT microswitch to turn on and off a double-pole relay, and then use the two pairs of contacts on the relay to operate the clock and the light.

Problem. Draw a circuit diagram for a circuit in which the release in pressure on a SPDT microswitch causes a lightbulb to light and a clock to start. Then build the circuit, substituting two flashlight bulbs for the bulb and the clock.

Solution. There are many solutions to this problem. The simplest is to connect the bulb and clock in series across the microswitch. This solution is one theoretically, but it is not very practical. First of all, most clocks require 115 volts a-c to run them, as do most light bulbs. In order to run the two in series, 220 volts a-c are needed, a voltage not usually available. Second, most clocks require only a very small current, whereas most light bulbs require a relatively large one. Since the current through any two elements in series with each other must be the same in each element, either the light bulb would be very dim or the clock would burn.

A second easy solution is where the bulb and the clock are in parallel. This is a practical solution only if the clock and the bulb happen to have the same rated voltage (e.g., 115 volts a-c). The series and parallel solutions are thus applicable only in special cases. To illustrate the problem in its general form, assume that the clock runs on 115 volts and the bulb on 6 volts. Under these conditions, neither the series nor the parallel solution will work, and a different type of solution, requiring a relay, is suggested.

Notice that this diagram contains three more or less separate subcircuits. Actually, there are no *electric* connections between any of these three subcircuits, but there are mechanical and magnetic connections. Look at them one at a time. Subcircuit *A* is a circuit in which the microswitch, the relay, and a source of power are in series. Releasing the pressure on the microswitch causes its normally closed contacts to close, allowing current to flow through the coil of the relay, pulling in the relay armature. Subcircuit *B* contains a bulb, a power source, and a pair of normally open contacts, all in series.

When the contacts close, the bulb will light. The contacts are mechanically connected to the relay so that, when the relay armature pulls in, they close. When the microswitch pressure is released, current flows through the relay coil, the armature is pulled in, the contacts close, and the light goes on. Subcircuit *C* allows the clock to start at the same time as the bulb lights, by using a second pair of normally open contacts on the same relay.

Now try to draw the circuit diagram just discussed without looking back at it. An efficient procedure is to begin by drawing the symbols for each of the elements that you know must be in the circuit, namely, a microswitch, a relay coil, a clock, and a light bulb. Then begin to connect them up so that each operates properly. Each of the separate subcircuits is drawn with its own power supply.

This is a convenient way to go about designing the circuit. When the circuit is actually built, however, it may turn out that the same power supply will be used in several subcircuits; in this case, the final circuit diagram should show that this is true (cases like this will be illustrated and discussed later). But in the initial stages of designing any circuit, it is much easier to represent a separate power supply for each circuit element that requires power.

However, the problem also includes the building of such a circuit, substituting two flashlight bulbs for the clock and the light. That requires some changes, not in the logical but in the practical aspects of the circuit diagram.

First of all, there is only one relay on your board that has enough contacts on it and the rating of the coil of that relay is 115 volts a-c. Therefore, a new symbol for 115 volts a-c must be substituted for battery 1. The two light bulbs both require the same voltage, and they might as well be run off the same battery, especially since you only have one battery. The circuit diagram should be changed to indicate this. Connect up this circuit on your board.

Here, for the first time, you must be careful because, when the circuit is plugged in, there will be enough voltage across the relay and the switch to give a good shock. In general, never touch any terminals or bare wires when the circuit is plugged into the wall socket.

A convenient way to bring 115 volts a-c to your circuit board is as follows. Connect the wall plug to one end of your lamp cord. Then, at the other end of the cord, solder the two conducting paths to a pair of terminals on one of the terminal strips mounted on your board. Thereafter, when the plug is plugged into the wall, the two terminals may be regarded as the source of 115 volts a-c.

Caution: Finish the circuit and check the connections against the diagram before you plug it in. Then, as a final check, measure the resistance between the two prongs of the wall plug with an ohmmeter (again, *before* it is plugged into the wall). The ohmmeter should show an infinite resistance when the switch is in the position where the relay is off, and a resistance of at least a few hundred ohms (the resistance of the relay coil) when the switch is in the other position. This test is a good one to use generally when you are about to plug in a new circuit. If you have made a wiring mistake which causes the resistance across the plug to be very low (e.g., 1 ohm), and you were to plug the circuit into the wall, a very large current would flow (e.g., 115 amperes), and a fuse would blow out.

Another convenient way to check a new circuit is to connect a large light bulb in series between the circuit and the wall socket. The bulb should be of such a wattage that, in order for it to light, it must conduct a current larger than that normally drawn by the circuit to be tested (e.g., if the circuit is supposed to draw I ampere, the bulb should be 500 watts or more). If the circuit has been correctly wired and then plugged in, the bulb will not light and will just add a small resistance to the circuit. However, if there is a short circuit in the wiring, the bulb will light fully and no damage will be done.

Problems

- Given a single 3-volt battery, a 3-volt motor that reverses direction of rotation when the direction of current through it is reversed (e.g., the motor on your circuit board), and a DPDT switch, design a circuit to drive the motor either clockwise or counterclockwise, depending on the position of the switch.
- Given two bulbs, *A* and *B*, design a circuit which lights *A* when a selector switch is in one position; *B* when in another position; and *A* and *B* when in a third position.
- Given two bulbs, *A* and *B*, and three push buttons, design a circuit so that pushing one button will light *A*, another *B*, and the third *A* and B.
- Given a Skinner box with two adjacent bars, design a circuit so that the experimenter will have a selector switch which allows him to reinforce the pressing of (a) bar *A* alone (i.e., *A* and *B* together will not be reinforced); or (b) bar *B* alone; or (c) bars *A* and *B* simultaneously.
- Design a circuit to measure choice reaction time. The experimenter closes his key, which lights either bulb *A* or bulb *B*, and starts the clock. The subject has two keys and presses one if *A* lights, the other if *B* lights. The subject's keys, when pressed, stop the clock. The subject holds his key down until he is told to release it.

- Design the circuit so that any response, correct or incorrect, will stop the clock.
- Design it so that only a correct response stops the clock.

- Design and build a circuit, using the components on your board, which will cause a 6-volt bulb to go on and the 3-volt motor to go off, when the shaft of the 5000 ohm variable resistor is turned through about 90 degrees.
- A subject is presented with three toggle switches, *A*, *B*, and *C*, and a push button. He is to learn which pair of switches is correct for each of a number of different stimuli (e.g., when he is shown a square, he is to turn on *A* and *B*, when shown a circle, *B* and *C*, etc.). He is shown a figure, throws the two switches he thinks are correct, then pushes the button. If he was correct, the button lights a bulb, and if wrong, it sounds a buzzer. Design the circuit so that the experimenter can select the pair of "correct" switches before each trial, and the rest is automatic.

COMPLEX SWITCHING CIRCUITS

All of the subsequent examples and problems will be typical of those which come up repeatedly in research. Many of the examples deal with learning experiments on animals because complex switching circuitry is even more extensively used in research on rats and pigeons in Skinner boxes than in the study of human behaviour. Probably this is a result of the fact that by verbal instruction it is possible to throw into operation a human subject's own built-in, logical machinery, thus replacing a lot of relays.

This chapter contains many problems, and it is important to solve them all. Solving a problem means drawing the solution on real paper. It often *seems* possible to read a problem and arrive at a solution without actually writing anything down. However, thinking up a way of solving a problem is easy. Solving it is the hard part.

Problem. John has a ball that weighs 3 pounds. It is hollow, and made of rubber weighing 63 pounds per cubic foot. The ball just barely floats in water. How thick are its walls?

Solution. Use division.

HOLDING RELAYS

There is one particular relay circuit that is used very extensively. It is called the holding, or lockup, circuit, and is used when a circuit element must remain activated after the activating circuit is turned off. In essence, a relay is made to close as soon as a switch is operated, hold itself closed even after the switch is opened, and release only when a second switch is operated. The basic circuit is illustrated in the following example.

Suppose you are given the problem of designing a circuit to measure the length of time it takes a rat to run from point *X* in a straight alley to another point *Y* farther down the alley. First, some device must be used to sense when the rat passes each of the two points. In solving the present problem, we will use the method mentioned in the previous chapter of hinging a section of the floor at *X* and another at *Y*, so that the rat's weight will momentarily operate a SPDT microswitch at each of the two points. The clock commonly used for this type of timing consists of an electric motor driving the hands through an electric clutch.

The motor runs constantly, but the hands turn only so long as the clutch is engaged. The clutch operates as long as a small current is being passed through it, and disengages as soon as the current is turned off. To time a rat, then, a circuit is needed which will cause a current to begin to flow through the clutch when one SPDT switch (at *X*) is momentarily operated, and stop fl owing when another (at *Y*) is operated. Neither microswitch alone could operate the clutch in the desired way because, during the time the rat is between *X* and *Y*, both switches are open so that the clutch would not be engaged.

Suppose, instead, that the clutch is turned on and off through a relay. Now, if the relay could be made to close when switch *X* is closed, hold itself closed even after switch *X* is opened, and release when switch *Y* is operated, the problem would be almost solved. (The clutch could then be turned on and off by a pair of contacts on the relay.) The clutch connected through a relay, and the relay connected directly to one of the microswitches. This circuit is no improvement over connecting the clutch directly through the microswitch.

As soon as *X* is closed by the rat's running over that section of the alley, the relay closes, allowing current to flow in the clutch circuit. However, as soon as the rat leaves that section of the alley, the microswitch will open again, the relay will thus be released, and the clutch is turned off. These contacts are connected in parallel with the normally open contacts of the microswitch. The action of the circuit is as follows. At first, the microswitch and the relay are open so that no current flows anywhere (except, of course, through the clock motor). (Be sure that it is clear from the diagram that no current can flow. Try to find a closed path from one side of a power source to the other side of the same source. If there is no such path, then there can be no current flow, but as soon as there is such a path, there must be current flow.)

Now the rat steps on the hinged alley section, closing the normally open contacts of the microswitch. This furnishes a path for current to flow through the relay coil (dashed arrows), and the resulting magnetic field will cause the relay armature to pull down, closing the two pairs of normally open contacts. The top pair activates the clutch. The lower pair provides a second path for the current to flow through the relay coil (dotted arrows). This time, when the rat leaves the section and the microswitch opens again, the relay will stay closed because there is still a path allowing current to flow through the coil. With the microswitch open, current will flow only along the dotted lines. Under these conditions, the relay is, in a sense, holding itself closed and once it has been closed, microswitch *X* no longer has any effect at all.

The clock will start running when the rat first crosses the hinged section of the alley and will keep running indefinitely, as the circuit is now drawn. Even if the rat were to retrace and go back and forth several times over this alley section, the clock would be unaffected.

The relay is your 115-volt a–c relay, and switch X is one of the microswitches. (Observe the same precautions as were discussed in the previous chapter: remember that 115 volts can be painful.) If the circuit is connected up correctly, the light should go on when you first press the microswitch and stay on after you let go. Now the bulb is lit (the clock is running). How do you turn it off? There are several ways. First, you can unscrew the bulb.

This is the equivalent of putting a switch in the clutch circuit and opening it. Suppose that this switch were actually the second microswitch *Y* in the alley, connected. That is, the normally closed pair is connected in series with the clutch. When the rat hits section *X*, the relay closes and the clock starts. When he hits section *Y*, the clock stops. If at that moment the clock could be read very quickly, the circuit problem would be solved. But as soon as the rat leaves section *Y*, the clock will start up again. Prove this by unscrewing the bulb and then putting it back again.

If the section *Y* switch could actually turn off the power to the *relay*, then the relay would stay open even after the section is returned to its normally closed condition, because the only things keeping the relay closed are its own contacts. In the circuit on your board, turn off the light by pulling out the wall plug. Now when the plug is connected again, the light will remain off and the whole circuit is restored to its initial condition.

On your board, connect up the normally closed contacts of the second microswitch. This switch will now do what you were doing when you momentarily disconnected the wall plug. If the circuit is working properly, the light will turn on when switch *X* is momentarily pressed, and stay on until switch *Y* is momentarily pressed.

This circuit, the holding relay circuit, is very fundamental to switching circuit design. It is used as it has been presented here or with only slight modifications in a very large proportion of the switching circuits found in behavioural research. For this reason, it is worthwhile for you to draw the circuit diagram over and over again until it becomes almost automatic. Be sure you understand its principles of operation. This will save you considerable time when you are solving many of the remaining problems in this book.

If the circuit were used in a straight alley way, it would permit the measurement of the total time between the first crossing of *X* and the first crossing of *Y* (i.e., stepping on *Y* again would not restart the clock). Consider instead the problem of designing a circuit to measure the total time the animal spends in the portion of the alley between *X* and *Y*. In other words, if he runs out of this portion, past *Y*, and then turns around and runs back into it once more, the clock should start again.

Since this is probably the most difficult problem yet presented, it may be a good one with which to illustrate the sort of process used by people skilled at circuit design to solve relatively complicated problems.

Many of the processes involved in arriving at solutions to this kind of problem are difficult to verbalize in a coherent way. If prodded, people who are skilled at circuit design frequently will say that they knew by intuition where to start and what things to try. Intuition is probably an excellent word to describe what happens. Most people using such intuition will readily agree that it is developed through a combination of some intelligence and a lot of experience. Here, an attempt will be made to enumerate some of the submerged logic involved.

The first step in arriving at a solution is to state the circuit requirements as explicitly as possible. In this particular problem, the circuit must:

- Start the clock any time the animal (*a*) crosses *X* going toward *Y* or (*b*) crosses *Y* going toward *X*.
- Stop the clock any time he (*a*) crosses *X* going away from *Y* or (*b*) crosses *Y* going away from *X*.

(For any problem there are a number of solutions. For example, this problem may be easily solved, in theory at least, by hinging the entire section between *X* and *Y* and resting it on a microswitch. In practice, it is difficult to make a large section of alley operate that way. The solutions presented herein are chosen both because they are practical and because they illustrate various important aspects of circuit design. If you arrive at a solution different from one in the text-and this happens very often—that does not necessarily mean that yours is incorrect.)

Examination of the apparatus requirements listed above indicates that the circuit must sense the direction in which the rat is traveling as well as his presence when he is at point *X* or point *Y*. If there were two short, hinged floor sections at *X*, and if the circuit could determine the order in which these two sections were traversed, it would indicate which way the animal was going when he went by *X*, and either of the sections might also indicate his presence at *X*. In other words, it may be that if there were four hinged floor sections, each with a microswitch, two at *X* and two more at *Y*, it would be possible to solve the problem. A guess about what kinds of circuit elements might solve the problem can be considered the second step in the solution.

The third step might then be to see whether or not such a collection of elements really can be made to work. To do this, it is probably most efficient to begin by representing, on a large sheet of paper, each of the possible elements without any particular regard for how they will be connected up.

The clock will run so long as there is current flowing through its clutch coil, and it becomes clear and considering the original problem that the clutch must be operated through something that is not yet represented. This is true because, although the clutch must remain activated when the rat is, say, halfway between *X* and *Y*, there is nothing drawn yet that will act that way. As in the preceding problem, a holding relay circuit is strongly indicated. A relay is thus drawn in with at least enough contacts to hold itself, and connected so as to operate the clutch. Now the clutch no longer need be considered. The problem is to get the relay to be on all the time the rat is between *X* and *Y*. If this is done, the clock will necessarily do the same.

Figuring out additional connections requires consideration of the step-by-step operation of the circuit. Let the rat come out of the start box and begin to walk up the alley. The clutch is to engage, that is, the relay is to close, when the rat passes *X*. Therefore, either microswitch X_1, or X_2 must close the relay. Try X_1 first. With this circuit, when the rat walks past *X* and stands between *X* and *Y*, the clock keeps running (the relay is connected to hold itself on). Now, if the rat turns around and crosses *X* again, the clock are supposed to go off. Since X_2 is left over, try its normally closed contacts to turn off the clock.

This will stop the clock all right when the rat recrosses X_2 going toward the start box, but two other things are wrong. First, if he goes a little farther and crosses X_1 too, the clock will go back on again. Second, after he leaves the start box in the first place, he will start the clock at X_1, but stop it again as he goes by X_2. Therefore, try reversing the roles of X_1 and X_2 to see if that works any better. Now, when the rat crosses X_1 going toward Y, nothing happens. When he crosses X_2, the clock starts. If he then turns around and recrosses X_2 going the wrong way, nothing happens.

But as soon as he crosses X_1 going the wrong way, the clock stops. When he again turns around and recrosses X going toward Y, the clock starts once more. So far, then, this circuit is operating according to plan. All that remains is to make the switches at Y operate in a similar, but *opposite*, manner. In other words, it would seem plausible, as a first guess, to make Y1 start the clock and Y_2 stop it. (Crossing Y toward X should do the same thing as crossing X toward Y.) The circuit is connected this way and will solve the problem. In this circuit, either Y_1 or X_2 will start the clock, and either X_1 or Y_2 will stop it.

The clock will always be running when the rat is between X and Y because the rat must have most recently closed either X_2 or Y_1 to get there. And any time the rat leaves the section between X and Y, the clock must be stopped because the rat must have most recently opened either X_1 or Y_2. The next problem illustrates another modification of a holding relay circuit. A relay that is turned on and off repeatedly at a fairly fast rate will buzz. In the holding relay circuit, the relay holds itself on. If instead a relay is connected in such a way that it turns itself *off*, then it must buzz.

Before the switch is closed, the relay is open, so that the normally closed contacts of the relay are closed. As soon as the switch is closed, current will flow through the relay contacts and through the relay coil, causing the relay armature to pull in. But when the armature pulls in far enough to open the normally closed contacts, the circuit through the relay coil is broken, so the armature is released. This allows the contacts to close again, and the cycle automatically repeats itself.

Connect up this circuit on your board, using the sensitive relay (not the 115-volt one). Begin by using as low a voltage across the relay as is convenient (say 3 volts), and increase it, if necessary, until the relay buzzes. Be sure that you understand which terminals are connected to which parts of the relay (i.e., which terminals on the body of the relay are internally connected to the relay coil) before you connect up the battery. If you cannot tell about some terminal by looking at the relay, check it with an ohmmeter to see to which part it is connected. Practically all buzzers commercially available operate on exactly the same principle as the one you have constructed. The only important difference between your buzzer and a doorbell buzzer is that the latter uses a much cheaper relay.

PRINCIPLES TO FOLLOW IN DESIGNING CIRCUITS

There are three principles which an experienced circuit designer uses automatically as he designs a new circuit. These principles are also useful in checking a novice's diagram, and you should apply them to each of your circuits when they are in the design stage as well as after you think the circuit diagram is complete.

PRINCIPLE 1

Every possible path of current from one side of each power source to the other side of the same source should contain some resistance.

Trace each such path. Each path should contain some circuit element which has resistance (e.g., a relay coil). If a path contains only a pair of closed relay contacts, for example, an infinite current will be drawn from the power source. Each circuit should be checked in this way under all its conditions of normal operation. That is, if during circuit operation, a relay is sometimes open and sometimes closed, check all paths under both states. Remember that *any* one path of zero resistance is a short circuit, even if there are many other paths between a and b that do contain resistance.

PRINCIPLE 2

There should be no permanent path of zero resistance between the two ends of any circuit element. For example, if there is a direct connection between the two ends of a relay, the relay will be short circuited and cannot operate. In some circuits, an element is deliberately short-circuited during some phase of the operation of the circuit as a whole, but there are other phases during which the short circuit is removed. If it were always present, the short-circuited element could never perform any function.

PRINCIPLE 3

Every circuit element should be part of a path from one side of a power source to the other side of the same source during some part of the operation of the circuit. Any element never in a path from one side of the source to the other cannot have any function in the circuit because current can never flow through it. The resistor is superfluous. Since there is no path from one side of either power source, through the resistor, and back to the other side of the same power source, current will never flow through the resistor.

Therefore it serves no function and can be omitted without changing the action of the circuit. The principles listed above, if followed, will eliminate some of the mistakes which prevent a circuit from operating. The following is a list of principles of a different sort. Although it is not logically necessary to employ them, their use generally results in circuits that are simpler to design and construct.

- If two or more circuit elements require the same kind of power supply (e.g., 115 volts a–c), connect one side of each of them together and to one side of the power supply. This is an extremely useful principle. Whenever you are faced with a circuit problem in which either one or another of a set of circuit elements is to be activated, or each element requires the same voltage, the first connections drawn in the diagram should be these. For example, to draw the diagram for a circuit to light any one of four bulbs, the first step should be as drawn.
- If two or more subcircuits are *nowhere* connected together, it is permissible to make one connection between them anywhere in

the circuits. For example a connection may be made from any point in the circuit labeled I to any point in the circuit labeled 2. Such a connection will have no effect on either subcircuit. This principle is useful, for instance, in the case where there are two subcircuits, one of which is supposed to go on when the other is off, and vice versa. The subcircuits will operate in this way, through the contacts of a DPDT relay. But if the only available relay is a SPDT one (e.g., a sensitive relay), the two circuits may be connected together and to the common contact.

- If two or more subcircuits are nowhere connected together, and if they both require the same kind of power source, a single power source may be used for both.

Label the two sides of one source "+" and "–"; next label the two sides of the other power source "+" and "–" (with alternating current, either side can be called "+"); then connect the "+'s" together and to one side of a single power supply, and the "–'s" together and to the other side of the power supply.

If there are several subcircuits to be connected in this way, the circuit diagram can be made easier to follow if the convention is followed. Here the points in each subcircuit into which power is to be led are labeled "+" and "–" and the power supply itself is represented, at the bottom of this figure, with its terminals labeled "+" and "–." When the circuit is connected up, all the "+'s" are connected together, as are all the "–'s."4. If two or more subcircuits are already connected together at one point, and they all require the same kind of power supply, it *may* be possible to use a single supply for all. To find out, label one side of each separate supply "+" and the other "–."

Then find the point where the two circuits are already connected together. If the connection happens to be between "+" in one subcircuit and "+" in the other, simply connect the two "–'s" together and to the "-" side of a single power supply.

If the connection is between two "–'s" the "+'s" can be similarly connected together. If the connection is between a "+" in one subcircuit and a "–" in the other, just reverse the labels in one of the circuits, and connect the "+'s" together and the "–'s" together. These cases reduce to the situation described in the preceding paragraph.If the pre-existing connection between two subcircuits is not at one of these places, then try to change it so that it is. For example, the subcircuits are not joined in any of the ways just described.

But the subcircuit 1 can be redrawn without changing its action, and the remaining connections may then be made to use a single power supply. It is often but not always s possible to change subcircuits in this way, so that a single power supply may be used.5. Do not connect any two subcircuits together at more than one place, except when connecting them as described in Principles 3 and 4 above. If, at some stage in the designing of a circuit, you would like to connect two subcircuits together, but they are already connected at another point, figure out some way to avoid making the second connection. The most common reason for wanting to connect two circuits together is to make double use of an SPDT contact set. If you face this problem, and the two subcircuits are already connected together somewhere else, the best solution is usually to substitute a DPDT contact set for the SPDT set (e.g., add another relay to the circuit), and thus avoid making the second connection between the two subcircuits.

Problems

- A subject is presented with a light whose colour slowly changes from red through the spectrum to violet. He is to indicate at what times it appears to be pure yellow, pure green, and pure blue. He does this by momentarily pressing the first of three keys when it is yellow, the second when it is green, and the third when it is blue. The experimenter then reads these times on each of three clocks. The experimenter turns on the light and starts all three clocks by momentarily pressing a button. Design the circuit.
- Design a circuit to measure choice reaction time in the following way. The experimenter momentarily presses a button which turns on one of two lights and starts the clock. The subject responds by momentarily tapping the appropriate one of two keys. If he presses the correct key, the clock stops. If he presses the wrong one before he presses the correct one, a buzzer turns on and stays on until the experimenter turns it off. The clock always continues to run until the correct response is made.
- An animal is in a Skinner box that contains two bars. The experimenter has a switch with which he can establish that the animal will be reinforced either for pressing bar *A* and then bar *B*, or for pressing bar *B* and then bar *A*. The bars are too far apart to be pressed simultaneously. The reinforcer operates once each time its terminals are connected together, and the experimenter resets the apparatus after each reinforcement. Design the circuit.

OTHER RELAYS

Some of these will be discussed here. In most cases the functions of the more complex relays may be duplicated by the proper combination of simpler relays. For this reason, it is not essential to be acquainted with a variety of types. However, it is often much more convenient to buy one complex relay than to try to duplicate its action by combining several simpler ones.

MECHANICAL ASPECTS

The physical size of a relay is generally determined by the electrical requirements. For example, if the relay is to switch large currents, its contacts must be big. However, relays of any given electric capacity do come in a range of sizes so that, should it be important for a circuit to be small, it is possible to buy extra small relays.

The relays on your circuit board are exposed, i.e., you can see most of the parts. Many relays, however, are sealed within dustproof and moisture proof cans, with only the terminal posts exposed. The trouble with sealed relays is that, since it is not possible to see and get at the working parts, they are somewhat less adaptable than exposed relays. But there are certain situations in which sealed relays must be used, e.g., if the circuit is to be operated in a very dusty or humid environment.

In addition, there are certain types of relays, such as very fast-acting ones, which must be sealed in order to work properly. Many sealed relays and some that are not sealed are of the plug-in type. This means that the connections are all brought out to the prongs on a plug. To wire such a relay into a circuit, all the connections are made to an appropriate socket and then the relay is simply plugged in. This type of relay is particularly suited for circuits which get a lot of use. Should a relay go bad, it may be replaced very easily.

ELECTRICAL ASPECTS

The coil of any given relay is constructed so that the armature will pull in at some particular voltage and current. Relays of the type used for most switching operations are rated in terms of the voltage necessary for proper operation. For example, the coil of one of the relays on your board is rated at 115 volts a-c. Relays are readily available for use with either a–c or d–c at voltages between about 6 and 220 volts.

Relays that require small amounts of power to operate are usually rated in terms of their resistance and the current necessary to operate them. They are called sensitive relays and generally have a relatively simple contact arrangement, either SPST or SPDT. The sensitive relay on your board is SPDT. These relays are not often used to perform complex switching operations, but only to allow a small amount of power to control a larger one. One of the most sensitive of all relays has an SPDT set of contacts.

One of the contacts is mounted on the needle of a sensitive ammeter, and the others are fixed to the frame of the meter. As current passes through the coil of the meter, the needle moves and, at some current value, the contacts close. Usually the contacts are magnetic so that, once they touch, they hold a good contact. It is then necessary to open the contacts by some mechanical device. This type of relay may be made to operate on as little as a few microamperes and less than a millivolt. The meter-type relay is delicate and relatively expensive. When such small amounts of power must control large amounts, it is much more common to use some form of electronic amplification and a less sensitive relay. This frequently turns out to be a cheaper solution.

It is important to realize that there are two different kinds of ratings for any given relay—the coil rating and the contact rating. For example, a relay may have a coil rating of 115 volts a-c and a contact rating of 200 volts at 3 amperes. This means that the relay will close properly on 115 volts a–c, and that the contacts opened and closed by the armature can carry up to 3 amperes and 200 volts without getting too hot. There is no logically necessary relationship between these two types of ratings on any given relay. It does happen, however, that a general relationship exists for practical reasons. If a relay is to have a large contact current rating, it must have large contacts and they must be held together tightly.

The contacts must also be relatively far apart when they are open. All these factors necessitate a relatively strong magnetic pull to close the relay, and this in turn requires a relatively high current or voltage rating on the coil of the relay. On the other hand, for a relay to be very sensitive, i.e., to be able to close with a very small amount of power in its coil, the contacts must be very light, and cannot therefore carry large currents.

There is a type of relay that is an exception to this general rule. The mercury-wetted relay can be made very sensitive although the contacts are capable of carrying relatively large amounts of current. The contacts of this type of relay are coated with mercury. When a pair of contacts meets, the pools of mercury on each contact join to form a relatively large contact area. Mercury-wetted relays have another property that is very desirable under some conditions. When the armature of a normal relay is pulled in, or when the lever of a microswitch is pushed, the contacts bounce against each other, and the resulting contact is intermittent for a short time immediately after closure.

When mercury-wetted contacts are used, the mercury pool maintains electric contact during the bounce phase, and there is no period of intermittance. To maintain the mercury in a clean condition, mercury-wetted relays are mounted in sealed containers which are usually filled with an inert gas under pressure. The connections are brought out to a plug at the base of the container. (Most relays can be operated in any position, but mercury-wetted relays only operate properly when used base down.)

SWITCHING OPERATIONS

LATCHING RELAYS

The relays described so far operate rather simply, even though combinations of them can perform very complicated logical operations. However, there are several types of relays which operate in a more complicated way and allow the rest of the circuit to be simpler. These relays actually substitute mechanical for electrical logical operations, so that the machinery gets complicated and the circuit gets simpler. This is called a latching relay. It consists of two independent coils, two interlocking armatures, and one set of contacts. When current passes through coil *A*, the armature *A* pulls down, closing the contacts. The armature *B* mechanically catches on armature *A* and holds it in the closed position even after the current stops flowing in coil *A*.

When current passes through coil *B*, armature *B* pulls in, releasing armature *A*, which returns to its initial position and opens the contacts. In other words, the operation of a latching relay is similar to that of an ordinary relay wired in a holding circuit. Latching relays are more expensive and harder to obtain than ordinary relays, and have only one real advantage over the holding circuit. A latching relay does not require a continuous flow of current to keep it latched. If an apparatus were to be battery-operated, for example, the latching relay might be worth the extra expense because of reduced drain on the battery.

SEQUENCE RELAYS

There is a class of special relays called sequence relays or ratchet relays, in which successive bursts of current through a coil cause a set of contacts to step through a particular sequence of openings and closings. Generally, when the current through the coil is turned on, the relay cocks, and when the current then goes off, the relay operates and is ready for the next cocking. For example, suppose that the space bar on a typewriter is pulled down by a magnetic coil. When the current goes on, the bar pulls down, cocking the typewriter carriage mechanism.

Then, when the current is turned off, the bar is released and the carriage actually shifts over one space. If there were a cam attached to the carriage that closed a pair of contacts on even numbered spaces and opened them on odd numbered spaces, then successive bursts of current through the magnet coil would produce alternate openings and closings of the contacts.

Current through the coil pulls an armature down, and the armature pushes the toothed wheel (which looks like a gear) through a part of a turn. There is a set of cams rigidly attached to the same shaft as the toothed wheel, and the cams open and close the contacts. In the figure, the position of the cam is such that the upper pairs of contacts are closed. When current is turned on through the coil, the armature will advance the toothed wheel one tooth, the cam will rotate through part of a turn, and the contacts will open. When the current through the coil is then turned off, the contacts will remain open but the armature will return to its "ready" position.

The next time current flows through the coil, the cam will close the contacts again, and so on. This relay operates when current is turned on and cocks when current is turned off. This is different from the typewriter example discussed above. Relays operating either way are available. The sequence relay operates in a way similar to the latching relay, except that the on-off series in the sequence relay is produced by repeated bursts of current through the *same* coil (repeated closures of the same switch), whereas the latching relay operated from two different coils (and switches).

There are occasions when it is very desirable to have successive closings of a single switch produce successive reversals of a contact configuration. One interesting example of this is the binary counting circuit.

PROBLEM. Given a Skinner box with a bar-activated microswitch (pressing the bar closes a microswitch), design a system for reinforcing the animal on any of a number of fixed ratios. For example, the animal may get reinforced for every second press, or every fourth, etc.

Solution. First, we will choose to make available each of the following schedules: Reinforce on:

Every press; or
Every second press; or
Every fourth press; or
Every eighth press.

Assume that a device is available which delivers one reinforcement each time it is turned on. The principle to be used in this problem, and for any problem involving binary counting, is that two changes of state in the driving unit are required for each single change of state in the follower. In this case, consider the microswitch as the first driver, in that it will drive an alternating sequence relay. When the switch is closed, the relay contacts will close, but when the switch opens again, the relay contacts stay closed.

The next time the switch closes, the contacts open. Thus, the switch must close and open twice for each time the relay contacts close and open once. If this relay is connected so that, in turn, it drives a second sequence relay, then the second relay will operate once for each two operations of the first relay, and thus once for each *four* operations of the switch. The sequence of operations of the entire circuit is also shown in this figure.

Note that there is a unique combination of closed relay contacts for each count from one through eight, and then the sequence repeats. Thus, looking at the relay contacts themselves, it would be possible to say what the count was.

This is the principle underlying essentially all of the binary counting systems used in counting and computing equipment. A standard double-pole relay (relay 0), a reinforcer, and extra pairs of contacts have been added to sequence relays 1 and 2. The circuit is designed on the following basis. In order to fulfill the problem requirements, it is necessary to find some particular contact pattern that occurs on each press, another that only occurs on every other press, another every fourth, and another only every eighth press. (Relay 0 serves only to add another pair of contacts to the microswitch, i.e., if that switch were a double-pole one, relay 0 would be unnecessary.)

The circuit operates as follows. First, consider what happens when the selector switch is in the position marked 1:1. Each time the rat presses the bar, the switch closes, thus closing the relay. The relay, in turn, allows current to flow through the reinforcement mechanism, delivering reinforcement at a ratio of one per press. Now let the selector switch be in position 1:2, as it happens to be in the circuit diagram. In this case, current can only flow through the reinforcer if the contacts on *both* relay 0 *and* relay 1 are closed, and from the table it can be seen that this happens every other time the bar is pressed. Similarly, relay 0, relay 1, *and* relay 2 must be closed for reinforcement when the selector is in the 1:4 position, and this occurs only once out of each four bar presses.

Suppose, now, that each relay in the counting circuit above went through a sequence of ten positions instead of just two (on and off). A decade counting circuit could then be designed on identical principles. Relays like these are essentially electromagnetically controlled selector switches, and are generally called stepping relays, or steppers. Stepping relays may be purchased with as many as 100 steps and 6 decks; that is, 6 mechanically connected rotating arms and 6 sets of 100 contacts. Such a relay would be a 6P100T relay.

Successive pulses through the coil step the arms around, always in the same direction. The stepper is called an add-subtract stepper. Current through the left-hand coil moves the wheel one step counterclockwise, and current through the right-hand coil one step clockwise. The post mounted on the wheel opens the contacts when the wheel is in one particular position. Add-subtract relays are also constructed with a contact arm mounted on the wheel which moves across a set of stationary contacts.

It operates as follows. As the arm steps forward, it stretches the spring. This spring exerts a force which tends to return the arm to its starting position, but normally a ratchet prevents the arm from doing so. However, when a voltage is applied to the reset coil, the ratchet is pulled back and the arm swings all the way back to its starting position.

When the arm arrives at position "5," the reset coil will be activated, allowing the arm to begin its travel back to position "0." But as soon at it leaves position "5," current stops flowing through the reset coil, and the ratchet may catch the arm before it gets all the way back to "0." To insure positive resetting, resettable stepping switches usually have an additional set of contacts, such as those labeled "reset contacts". These contacts are normally closed, but are forced open by the reset contact actuator when the arm is in the "0" position.

When the arm arrives at position "5," a holding relay is actuated, which holds the resetting coil on until resetting is completed, completion being signalled by the opening of the reset contacts. In many parts of the country, pinball machines serve as an excellent source of miscellaneous relays. Out-of-date machines or those that no longer work very well can often be purchased for $15 or $20, and the parts they contain are worth at least ten times that amount. For example, consider the unit that indicates the number of free plays earned.

The number appearing in the replay window is one of a set of numbers printed on a drum, and the drum is mounted on the axle of a stepping relay. Each time a new replay is won, the relay advances one step, displaying a new number. Each time a game is played, the stepper moves one step backward, subtracting one from the total available replays. Thus the stepper must be an add-subtract stepper. At the owner's pleasure, all replays may be erased. This is done by pressing a switch that resets the stepping relay to zero. Therefore, the relay is an add-subtract, resetting relay.

There are other more or less complicated stepping relays in most pinball machines (e.g., to keep score), as well as many standard relays, light bulbs, motors, etc. Usually these elements operate on 24 or 48 volts a-c, and the transformer necessary to obtain this from the 115-volt wall socket can be found somewhere in the pinball machine.

Problems

- Given a 4P8T and a SP3T stepping relay, a push button, and two bulbs, programme a probability guessing stimulus sequence such that successive pressings of the button automatically turn on either one bulb or the other according to a predetermined schedule that is random for 24 presses and then repeats.
- A pigeon is trained to peck key A when a stimulus patch is lit and to peck key B when it is dark. The intensity of the patch is changed by changing the resistance in series with the lamp circuit

in a stepwise manner. The keys are then connected to the intensity contròl circuit in such a way that a peck on key A makes the patch one step dimmer and a peck on key B makes it brighter. The pigeon will then keep the patch at his threshold brightness. Design the circuit.

- To maintain the pigeon's performance in the apparatus of Problem 2, he must be reinforced. Add a device to the circuit which automatically turns off the stimulus light on every 20th, 40th, 60th, etc., peck on either key, and then reinforces the pigeon for the next peck if it is on key B (the correct response). After the peck (right or wrong), the sequence takes up where it left off. Furthermore, on every 10th, 30th, 50th, etc., peck, the patch is lighted to well above threshold, and the pigeon is reinforced if the next peck is to key A (the correct response). Then the sequence resumes again.
- Design an apparatus to reinforce a rat for double alternation in pressing two bars, i.e., one reinforcement for each AABB or BBAA sequence. He should not be reinforced for BABAA, etc., but AAAABB will be reinforced.

Chapter 8

Shock Circuits

Electric shock has been widely used to motivate experimental subjects in the behavioural sciences. Two areas of confusion are evident in a great many of the studies in which shock is used. First, there is a lack of understanding of the physical principles which govern the parameters of the shock. Because of this, a number of unwarranted conclusions have been drawn about behavioural principles. Second, and closely related, there are very few studies of the behavioural effects of shock itself.

For instance, there are several different sorts of shock that can be administered (e.g., constant current, constant voltage, sine wave, square wave, etc.) and, up to the time of this writing, there has not been a single study that unequivocally evaluates the relative qualities of any of the different kinds of shock. This chapter contains an analysis of the physical characteristics of electric shock and a discussion of techniques for constructing devices which deliver shock.

FUNDAMENTALS OF SHOCK

The floor of the box consists of a number of parallel metal rods held in place and separated by two strips of insulating material, usually plastic. When a rat stands on such a floor, it is very likely that his feet will touch several different bars. If a voltage were applied between each bar and every other one, current would pass through those parts of his anatomy which are bridging the gap between bars. In the simplest shocking grid (the set of bars is called a grid), the bars are connected together so that a particular voltage appears between each bar and both of its neighbors. This is accomplished by numbering all of the bars consecutively, and then connecting all of the even ones together and all of the odd ones together.

When the wire from the even-numbered bars is connected to one side of a battery and the wire from the odd-numbered ones to the other side, the battery voltage will appear between every adjacent pair of bars. As it stands, such a shocking system has a number of limitations that will be discussed later. However, its simplicity makes it useful for discussing certain aspects of electric shock which are fundamental to more complicated systems as well. To simplify the situation further, assume that the rat happens to be standing with both front paws resting on an odd bar and both hind paws on an even one.

Current from the battery passes through the junction between a bar and the rat's front feet, through the cornified layers of skin, through the tissues of his legs and trunk, and back out the other set of legs and skin. In a sense, these parts of the rat are all in series with each other so that the same current must flow through all of them. If the total resistance of the path through the rat is 250,000 ohms (a typical value), and if the voltage is 100 volts, then the current flowing through the rat will be:

$$\frac{E}{R} = \frac{100}{0.25 \times 10^6} \; \frac{E}{R} = \frac{100}{0.25 \times 10^6} = 400 \text{ microamperes}$$

An important question is, "How shocking are 400 microamperes?" Or should we ask, "How shocking are 100 volts?" Or, "How shocking is a combination of 100 volts and 400 microamperes?" Consider the voltage question first. Phrased more exactly it is, "How effective to the rat is a 100-volt battery between the grids?" The more dead skin and dirt there are between the bars and his sensitive tissues, the higher is the rat's resistance (the term "rat" now includes the dead skin and dirt). Thus, less current will flow through him.

The circuit schematizes this situation. As the resistance due to dead skin increases, the voltage across the sensitive tissue, that is, the voltage that actually affects the rat, is reduced. This means that even though the battery happens to show 100 volts between its terminals at all times, the voltage experienced by the rat will be less than 100 volts, and will vary from time to time and from rat to rat. For example, rats usually urinate when they are subjected to the stress of electric shock and, since urine is a good conductor of electricity, its absorption by the dead skin can produce radical fluctuations in the voltage affecting the rat, even when the battery voltage is constant.

CONSTANT-VOLTAGE SHOCKER

One thing can be said unequivocally about the shock voltage in the circuit. The voltage between the bars (across the whole animal including his dead skin) is constant. This is in contrast to the current flowing through the animal and the voltage actually affecting him, both of which frequently change in this circuit.

When a rat is being shocked through a circuit such, the current through him and the voltage across his sensitive tissues may fluctuate because of changes in the resistance of his skin, but even more drastic fluctuations occur whenever he moves around. Under certain conditions and shocks, a rat will freeze, but in most cases he starts to run or jump. Every time he makes the slightest movement, the pressure exerted on the grids by his feet must change. Consequently, the area of contact between his feet and the bars must change.

If he begins to jump up, his feet must push down harder on the bars and a larger area of skin will come in contact with the bars. The larger contact area means, in turn, that the total resistance of the rat to current flow between bars must decrease. These changes in resistance represent very large per cent changes in the total resistance, because the major part of the resistance of a rat on a grid is the resistance of the contacts between his skin and the bars. If an ohmmeter is connected between the bars of a grid on which a rat is walking, the resistance will be found to vary rapidly over at least a 1000% range.

In this kind of shock circuit, each change in over-all resistance is accompanied by a directly proportional change in the current running through the rat. The current may fluctuate wildly while the voltage between bars stays constant. In general, when a battery is connected directly between the grids, the circuit is said to deliver a constant-voltage shock, and it will be referred to as a constant-voltage circuit (recognizing the incorrect implications of that label). The reason for referring to the circuit in this manner will become clear as soon as nonconstant-voltage circuits are discussed.

CONSTANT-CURRENT SHOCKER

The relationship just described can be reversed so that the current stays virtually constant while the voltage fluctuates, if the shocker is connected. This is called a constant-current shocker. Only two things are different in this circuit: The battery voltage has been increased to 1000 volts and a 9-megohm resistor has been put in series with the rat and the battery. Suppose that the rat is standing still and has a resistance of 1 megohm. It can easily be shown by Ohm's law that there will be 100 microamperes flowing through the resistor and the rat, and that a voltage of 900 volts will appear across the resistor and 100 volts across the rat. Thus, so long as the rat is standing still, he is in exactly the same situation as with the constant-voltage shocker.

However, if he now should start to jump up and, in so doing, reduce his resistance to, say, 100,000 ohms, the current through him will change from 1000/(1,000,000 + 9,000,000) = 100 microamperes to 1000/(100,000 + 9,000,000) = 110 microamperes. A tenfold change in his resistance will produce only a 10% change in the current flowing through him for the following reason. The rat's resistance is a small part of the total resistance in the circuit. Therefore, relatively big changes in his resistance have only a small effect on the over-all circuit current. This circuit is thus called a constant-current circuit. Note that, although the current stays virtually constant through large changes in the rat's resistance, the voltage across the rat will change drastically. Therefore this circuit is really a constant-current, variable-voltage circuit. In general, any time the rat's resistance is a very large part of the total resistance in a shocking circuit, that circuit tends to give a constant voltage and changing current between the grids.

When the rat's resistance constitutes a relatively small proportion of the total circuit resistance, the circuit will deliver a relatively constant current and a varying voltage. (Since resistance, voltage, and current are the only three terms in Ohm's law, the only way to make a shocker that will hold *both* voltage and current constant is to hold the resistance of the rat constant too. This can be accomplished by strapping the electrodes directly and very firmly to the animal.)

MATCHED-IMPEDANCE SHOCKER

Several studies to be found in the literature have used a shock circuit with properties about halfway between the two already discussed. This circuit, called a matched-impedance circuit. Here, a resistance approximately in the range over which the rat's resistance will fluctuate is connected in series with the battery. As the animal moves, he undergoes less drastic changes in voltage than in a constant-current circuit, and less drastic changes in current than in a constant-voltage circuit. On the other hand, neither current nor voltage is held constant. This can be thought of, then, as a sort of compromise circuit.

It is called a matched-impedance circuit for a very simple reason. Just as the resistance of a rat or resistor is a measure of its ability to conduct a direct current, impedance is a meassure of its ability to pass any kind of current, direct or alternating. In any given object such as a rat or a resistor, the impedance, measured in ohms, is either equal to or greater than the resistance, and can be considered to be made up of resistance plus certain other properties to be discussed later. A matched-impedance circuit is one in which the impedance of the external resistor is roughly matched to the impedance of the rat.

Whereas it is not appropriate to go into great detail here about the physiological and psychological aspects of shock, it should be pointed out that the preceding discussion concerning the unspecifiable electrical aspects of shock is far from complete. Actually, it seems unlikely that the total current passing through or the voltage across a rat is a very good index of anything. Current *density* (amperes per square inch cross-sectional area) through any given portion of the anatomy must be more closely correlated with the physiological effect of shock. With a constant-current circuit, the rat can reduce the current density by gripping the bars tightly to increase contact area and by crouching down to increase his body's cross-sectional area. In this regard, the constant-voltage circuit tends to give a more constant current density than the constant-current circuit.

Considerations of this type, derived from the purely physical properties of shock, have very important correlates in the behaviour of the shocked animal. For instance, it would seem likely that a rat in a constant-current shocker would learn to crouch and freeze (because this reduces the current density) as well as not to jump off the grids (because at both the instant of leaving the grid and recontacting it, the areas of contact will be very small and the current density there very great). A rat in a constant-voltage shocker might learn the opposite pattern.

ALTERNATING-CURRENT SHOCK CIRCUITS

The previous discussion was based on shock circuits which use a battery as a voltage source. It is usually better, in practice as well as in theory, to use an a-c shock source. All of the foregoing principles apply to a-c as well as to d-c shocks, and since the properties of a-c circuits have not yet been fully described, d-c circuits were used in the preceding discussion. However, there are certain properties peculiar to a-c circuits which are relevant to shock boxes.

The animal, in this circuit, would get an a-c shock. Current would flow through him just as in the d-c circuit, except that the direction of current flow would keep reversing itself each time the switch is thrown. Now suppose you were to connect a zero-centre voltmeter across the grids. In a voltmeter, when the current flows one way, the pointer moves to the right. When it flows the other way, the pointer moves left. In the usual meter, a leftward movement cannot be read, but in a zero-centre meter it can. When the reversing switch is in one position, the meter will read + 115, and when it is reversed, the pointer will swing over until it reads – 115 volts.

A periodic repetition of the switching will result in the temporal voltage pattern. The a-c voltage represented by this figure would be described as a square-wave shock of 115-volt amplitude, with a fundamental frequency equal to half the number of reversals per second. Square-wave shocks are used extensively to stimulate nerve tissue in physiological preparations.

The shock from the circuit is a sine-wave shock, because the voltage delivered to wall sockets is sinosoidal. In almost all parts of the United States, the line power is held to a frequency of 60 cycles per second, and to a voltage of approximately 115 volts.

For the sine wave, the term "fundamental frequency" has the same meaning as for the square wave. It is the number of cycles occurring per second. For the sine wave, the "fundamental frequency" is the only frequency, and so the term "fundamental" is usually dropped. The square wave can be said to contain many (theoretically an infinity of) harmonic frequencies in addition to the fundamental, and for this reason the term fundamental is used in describing the number of complete cycles per second.

The simplest measure of voltage of a sine wave is the so-called peak to-trough amplitude. But there is another more common measure of voltage that requires some explanation. If two light bulbs were lit, one by a *square-wave* of 115 volts (230 volts peak-to-trough) and the other by direct current at 115 volts, the two bulbs would be equally bright because the direction of current flow through a light bulb does not affect its brightness. If the direct current were now left across one bulb and a very low frequency sine-wave voltage of 230 volts peak-to-trough put across the other, the one lit by the sine wave would always be either dimmer or, at the very top and bottom of the cycle, equally bright.

The average brightness thus would be less for the sine wave than for the direct current or for a square wave of equal peak-to trough voltage. If the frequency were increased to, say, 60 cycles per second, the changes in voltage would be so rapid that the bulb would not follow them; its brightness would remain fairly constant at a level lower than the brightness of the d-c-powered bulb. To light a bulb with a sine wave and make it just as bright as one lit by 115 volts d-c, the magnitude of the sine wave would have to be increased. It turns out that the necessary peak-to-trough voltage of the sine wave is $2 \times 230 = 324$ volts.

The power delivered by a sine wave of 324 volts, peak-to-trough, equals the power of direct current at 115 volts. Since it is generally more useful to specify power than peak-to trough voltage, a-c voltages are usually rated in terms of their equivalent d-c voltages. Therefore, to say that the wall socket voltage is 115 volts a-c means it will light a bulb with the same brightness as a 115-volt d–c source.

Its peak-to-trough voltage is about 324 volts. Whenever the voltage of a sine wave is given, it should be understood to mean the equivalent d–c voltage, not the peak-to-trough voltage. (This equivalent voltage is usually called the rms, or root-mean-square voltage, because it is actually the square root of the average of the square of the instantaneous voltages, integrated over one complete cycle.)

The shock circuit is a constant-voltage circuit and delivers a 115–volt, 60–cycle shock. For a lower voltage shock, a voltage divider may be added. This divider works in exactly the same way as it would for a d-c supply. To make the circuit deliver more than 115 volts is not so easy. If it were direct current we could just add some more batteries in series, but the only source of alternating current usually available is the wall socket, at 115 volts.

A simple and efficient device that gives higher or lower voltages than 115 volts is called the transformer. To make the circuit into a constant-current shocker, a very high resistance must be added in series with the source and the animal. Such a circuit definitely requires a transformer to obtain higher source voltages.

CONSIDERATIONS IN THE DESIGN OF SHOCK BOXES

GRID CONSTRUCTION

When an animal is disturbed, as by shock, he may defecate. The bars of a grid floor must be far enough apart for the feces to fall through. If they do not, they may rest between two bars and shortcircuit the shock to the animal. The bars should be of a material that does not corrode readily, because corrosion, helped along by urine, can form a layer that insulates the animal's feet from the bars.

When a rat has been shocked a few times, he tends to stand with his body oriented perpendicular to the direction of the bars. Therefore, if you want him to press a key at the end of the box to turn off the shock, for instance, be sure that the grid bars run across the box, not lengthwise.

SCRAMBLERS

In some types of training, a single shock is given in a pulse so short that the animal cannot move very much during the time it is actually on. However, in many applications, the shock is left on for an appreciable period, and the animal runs and jumps. In these cases, if the voltage is connected simply between odd and even numbered bars, the rat will soon learn to avoid shock by standing so that his feet are all on odd or on even numbered bars. This difficulty could be avoided if the grids were put so close together that each foot would have to overlap more than one bar, but then the grid would quickly become short circuited by feces. The solution is to introduce the voltage to the grids in a continuously changing pattern, in such a way that the animal cannot avoid shock by assuming any particular position. Devices that do this are called grid scramblers.

In this circuit, each time the stepping switch moves one step, the shock voltage is shifted to a new adjacent pair of bars. The stepper is driven continuously by the cam and microswitch, and the shock sweeps from one end of the grid to the other. The disadvantage of this simple scrambler is that the animal can avoid the shock altogether if he never touches two adjacent bars. The potentiometer *P* serves as an intensity adjustment. The motor *M* drives the contact arm *A* around so that it continuously sweeps the grids, connecting each in turn to the slider arm of the potentiometer.

All of the grids are also connected, through resistors, to one end of the potentiometer. In this way, the animal will be shocked at least once for each revolution of the contact arm, no matter what pair of bars he straddles. For example, suppose the rat is touching the bars labeled I and 2. During the time the slider is at any position other than I or 2, the rat will receive no shock, since there is no voltage between bars 1 and 2. However, when the slider arrives at bar 1 (which it does once during each revolution), the effective circuit will be as diagrammed.

The voltage will be applied across the rat and R_2 in series. When the slider contacts bar 2, the voltage will be applied across the rat and R_1 in series. Thus, no matter what pairs of bars he touches, he will be shocked twice for each revolution of the motor, at an intensity determined by the setting of the potentiometer and by the size of the resistors (100,000 ohms). The minimum shock intensity from this circuit is zero (when the slider on the potentiometer is all the way down), and the authors state that the maximum value attainable, with the slider all the way up) is a strong shock for a rat.

The scramblers described thus far have used stepping switches and rotating commutator arms to connect the shock between changing sets of grid bars. A number of scramblers have been developed which employ different means of switching (e.g., gas tubes, transistors, relays), but all of them are based upon the same principle, namely, that the shock voltage must be switched rapidly between different sets of bars.

Chapter 9

Amplifiers

The basic action of most of the electronic circuits used in behavioural research is one of amplification. That is, an electrical signal of low voltage and/or amperage is put in and the output is a more powerful signal, correlated in some way with the input. Such amplification is performed not by a process analogous to magnification, but by allowing a low power input to control the larger power of the output, just as the small effort required to open a valve may produce large changes in the amount of power coming out of a water pipe. (The electron tube is called a valve in England.)

Generally, an amplifier or simply amp, is any device that changes, usually increases, the amplitude of a signal. The "signal" is usually voltage or current. The relationship of the input to the output of an amplifier — usually expressed as a function of the input frequency — is called the transfer function of the amplifier, and the magnitude of the transfer function is termed the gain.

In popular use, the term usually refers to an electronic amplifier, often as in audio applications to operate a loudspeaker that is being used in a PA system to make the human voice louder or play recorded music. Amplifiers may be classified by the input (source) they are designed to amplify (such as a guitar amplifier to perform with an electric guitar), or named for the device they are intended to drive (such as a headphone amplifier), or by the frequency range of the signals (Audio, IF, RF and VHF amplifiers for example), or grouped by whether they invert the signal (inverting amplifiers and non-inverting amplifiers, or by the types of device used in the amplification (valve or tube amplifiers, FET amplifiers, etc.).

A related device that emphasizes conversion of signals of one type to another (for example, a light signal in photons to a DC signal in amperes) is a transducer, a transformer, or a sensor. However, none of these amplify power. The quality of an amplifier can be characterized by a number of specifications, listed below.

Gain

The gain of an amplifier is the ratio of output to input power or amplitude, and is usually measured in decibels. (When measured in decibels it is logarithmically related to the power ratio: $G(\mathrm{dB})=10\log(P_{out}/(P_{in}))$. RF amplifiers are often specified in terms of the maximum power gain obtainable, while the voltage gain of audio amplifiers and instrumentation amplifiers will be more often specified (since the amplifier's input impedance will often be much higher than the source impedance, and the load impedance higher than the amplifier's output impedance).

Example: An audio amplifier with a gain given as 20dB will have a *voltage gain* of ten (but a power gain of 100 would only occur in the unlikely event the input and output impedances were identical).

Bandwidth

The bandwidth (BW) of an amplifier is the range of frequencies for which the amplifier gives "satisfactory performance". The "satisfactory performance" may be different for different applications. However, a common and well-accepted metric are the half power points (i.e. frequency where the power goes down by half its peak value) on the power vs. frequency curve. Therefore bandwidth can be defined as the difference between the lower and upper half power points. This is therefore also known as the "3 dB bandwidth. Bandwidths (otherwise called "frequency responses") for other response tolerances are sometimes quoted ("1 dB, "6 dB etc.) or "plus of minus 1dB" (roughly the sound level difference people usually can detect).

A full-range audio amplifier will be essentially flat between 20 Hz to about 20 kHz (the range of normal human hearing). In minimalist amplifier design, the amp's usable frequency response needs to extend considerably beyond this (one or more octaves either side) and typically a good minimalist amplifier will have "3 dB points < 10 and > 65 kHz.

Professional touring amplifiers often have input and/or output filtering to sharply limit frequency response beyond 20 Hz-20 kHz; too much of the amplifier's potential output power would otherwise be wasted on infrasonic and ultrasonic frequencies, and the danger of AM radio interference would increase. Modern switching amplifiers need steep low pass filtering at the output to get rid of high frequency switching noise and harmonics.

Efficiency

Efficiency is a measure of how much of the input power is usefully applied to the amplifier's output. Class A amplifiers are very inefficient, in the range of 10–20% with a max efficiency of 25%. Class B amplifiers have a very high efficiency but are impractical because of high levels of distortion. In practical design, the result of a tradeoff is the class AB design.

Modern Class AB amps are commonly between 35–55% efficient with a theoretical maximum of 78.5%.

Commercially available Class D switching amplifiers have reported efficiencies as high as 90%. Amplifiers of Class C-F are usually known to be very high efficiency amplifiers. The efficiency of the amplifier limits the amount of total power output that is usefully available. Note that more efficient amplifiers run much cooler, and often do not need any cooling fans even in multi-kilowatt designs. The reason for this is that the loss of efficiency produces heat as a by-product of the energy lost during the conversion of power. In more efficient amplifiers there is less loss of energy so in turn less heat.

Linearity

An ideal amplifier would be a totally linear device, but real amplifiers are only linear within certain practical limits. When the signal drive to the amplifier is increased, the output also increases until a point is reached where some part of the amplifier becomes saturated and cannot produce any more output; this is called clipping, and results in distortion.

Some amplifiers are designed to handle this in a controlled way which causes a reduction in gain to take place instead of excessive distortion; the result is a *compression* effect, which (if the amplifier is an audio amplifier) will sound much less unpleasant to the ear. For these amplifiers, the 1 dB compression point is defined as the input power (or output power) where the gain is 1 dB less than the small signal gain.

Linearization is an emergent field, and there are many techniques, such as feedforward, predistortion, postdistortion, EER, LINC, CALLUM, cartesian feedback, etc., in order to avoid the undesired effects of the non-linearities.

Noise

This is a measure of how much noise is introduced in the amplification process. Noise is an undesirable but inevitable product of the electronic devices and components. The metric for noise performance of a circuit is Noise Factor. Noise Factor is the ratio of input signal to that of the output signal.

Output Dynamic Range

Output dynamic range is the range, usually given in dB, between the smallest and largest useful output levels. The lowest useful level is limited by output noise, while the largest is limited most often by distortion. The ratio of these two is quoted as the amplifier dynamic range. More precisely, if S = maximal allowed signal power and N = noise power, the dynamic range DR is $DR = (S + N)/N$.

Slew Rate

Slew rate is the maximum rate of change of output variable, usually quoted in volts per second (or microsecond). Many amplifiers are ultimately slew rate limited (typically by the impedance of a drive current having to overcome capacitive effects at some point in the circuit), which may limit the full power bandwidth to frequencies well below the amplifier's small-signal frequency response.

Rise Time

The rise time, t_r, of an amplifier is the time taken for the output to change from 10% to 90% of its final level when driven by a step input. For a Gaussian response system (or a simple RC roll off), the rise time is approximated by:

t_r * BW = 0.35, where t_r is rise time in seconds and BW is bandwidth in Hz.

Settling Time And Ringing

Time taken for output to settle to within a certain percentage of the final value (say 0.1%). This is called the settle time, and is usually specified for oscilloscope vertical amplifiers and high accuracy measurement systems. Ringing refers to an output that cycles above and below its final value, leading to a delay in reaching final value quantified by the settling time above.

Overshoot

In response to a step input, the overshoot is the amount the output exceeds its final, steady-state value.

Stability Factor

Stability is a major concern in RF and microwave amplifiers. The degree of an amplifiers stability can be quantified by a so-called stability factor. There are several different stability factors, such as the Stern stability factor and the Linvil stability factor, which specify a condition that must be met for the absolute stability of an amplifier in terms of its two-port parameters.

Electronic Amplifiers

There are many types of electronic amplifiers, commonly used in radio and television transmitters and receivers, high-fidelity ("hi-fi") stereo equipment, microcomputers and other electronic digital equipment, and guitar and other instrument amplifiers. Critical components include active devices, such as vacuum tubes or transistors. A brief introduction to the many types of electronic amplifier follows.

The term amplifier as used in this article can mean either a circuit (or stage) using a single active device or a complete system such as a packaged audio hi-fi amplifier.

An electronic amplifier is a device for increasing the power and/or amplitude of a signal. It does this by taking energy from a power supply and controlling the output to match the input signal shape but with a larger amplitude. In this sense, an amplifier may be considered as modulating the output of the power supply.

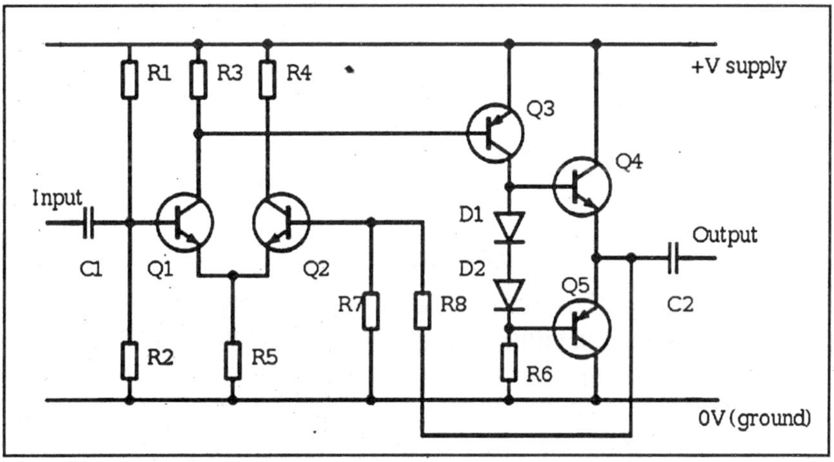

Fig. Types of Amplifier

Amplifiers can be specified according to their input and output properties. They have some kind of gain, or multiplication factor relating the magnitude of the output signal to the input signal. The gain may be specified as "output voltage/input voltage", "output power/input power" or any other combination of current, voltage and power. In many cases, with input and output in the same units, gain will be unitless; for others this is not necessarily so – for example, a transconductance amplifier has a gain with units of conductance (output current per input voltage).

In most cases an amplifier should be linear, that is the gain should be constant for any combination of input and output signal. If the gain is not linear, e.g. by clipping the output signal at the limits of its capabilities, the output signal is distorted.

Classification of Amplifier Stages and Systems

There are many alternative classifications that address different aspects of amplifier designs, and they all express some particular perspective relating the design parameters to the objectives of the circuit. Amplifier design is always a compromise of numerous factors, such as cost, power consumption, real-world device imperfections, and a multitude of performance specifications. Below are several different approaches to classification:

Input and output variables

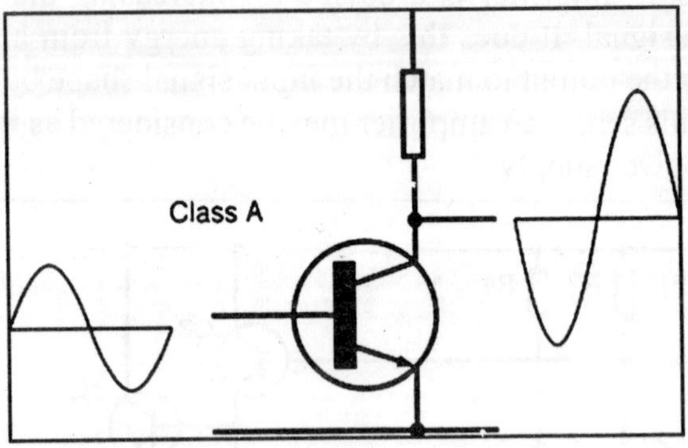

Fig. The Four Types of Dependent Source; control Variable on left, output Variable on Right

Electronic amplifiers use two variables: current and voltage. Either can be used as input, and either as output leading to four types of amplifiers. In practice the ideal impedances are only approximated. For any particular circuit, a small-signal analysis often is used to find the impedance actually achieved. A small-signal AC test current I_x is applied to the input or output node, all external sources are set to zero, and the corresponding alternating voltage V_x across the test current source determines the impedance seen at that node as $R = V_x/I_x$.

Amplifiers designed to attach to a transmission line at input and/or output, especially RF amplifiers, do not fit into this classification approach. Rather than dealing with voltage or current individually, they ideally couple with an input and/or output impedance matched to the transmission line impedance, that is, match *ratios* of voltage to current. Many real RF amplifiers come close to this ideal. Although, for a given appropriate source and load impedance, RF amplifiers can be characterized as amplifying voltage or current, they fundamentally are amplifying power.

Common Terminal

One set of classifications for amplifiers is based on which device terminal is common to both the input and the output circuit. In the case of bipolar junction transistors, the three classes are common emitter, common base, and common collector. For field-effect transistors, the corresponding configurations are common source, common gate, and common drain; for triode vacuum devices, common cathode, common grid, and common plate.

Unilateral or Bilateral

When an amplifier has an output that exhibits no feedback to its input side, it is called unilateral. One consequence is the amplifier has an input impedance that is independent of the load attached to the amplifier, and an output impedance that is independent of the signal source driving the amplifier.

The opposite case is the bilateral amplifier, where feedback connects the output to the input side of the amplifier. Such feedback often is deliberate, for example negative feedback often is used to tailor amplifier behaviour. However, at least as often, feedback is both undesirable and unavoidable; introduced, for example, by parasitic elements like inherent, undesirable capacitances in transistors that couple input to output. In any case, a bilateral amplifier has an input impedance that depends upon the load attached to the amplifier, and an output impedance that depends on the source driving the amplifier.

Linear unilateral and bilateral amplifiers can be represented by two-port networks. Most amplifiers are bilateral to some degree, however they may often be modeled as unilateral under certain operating conditions to simplify the analysis.

Inverting or Non-Inverting

Another way to classify amps is the phase relationship of the input signal to the output signal. An inverting amplifier produces an output 180 degrees out of phase with the input signal (that is, an inversion or mirror image of the input as seen on an oscilloscope). A non-inverting amplifier maintains the phase of the input signal waveforms. An emitter follower is a type of non-inverting amplifier, indicating that the signal at the emitter of a transistor is following (that is, matching with unity gain but perhaps an offset) the input signal. This description can apply to a single stage of an amplifier, or to a complete amplifier system.

Function

Other amplifiers may be classified by their function or output characteristics. These functional descriptions usually apply to complete amplifier systems or sub-systems and rarely to individual stages.

- A servo amplifier indicates an integrated feedback loop to actively control the output at some desired level. A DC servo indicates use at frequencies down to DC levels, where the rapid fluctuations of an audio or RF signal do not occur. These are often used in mechanical actuators, or devices such as DC motors that must maintain a constant speed or torque. An AC servo amp can do this for some ac motors.

- A linear amplifier responds to different frequency components independently, and does not generate harmonic distortion or intermodulation distortion. A nonlinear amplifier does generate distortion.
- A wideband amplifier has a precise amplification factor over a wide range of frequencies, and is often used to boost signals for relay in communications systems. A narrowband amp is made to amplify only a specific narrow range of frequencies, to the exclusion of other frequencies.
- An RF amplifier refers to an amplifier designed for use in the radio frequency range of the electromagnetic spectrum, and is often used to increase the sensitivity of a receiver or the output power of a transmitter.
- An audio amplifier is designed for use in reproducing audio frequencies. This category subdivides into small signal amplification, and power amps which are optimised for driving speakers, sometimes with multiple amps grouped together as separate or bridgeable channels to accommodate different audio reproduction requirements.
- A special type of amplifier is widely used in instruments and for signal processing, among many other varied uses. These are known as operational amplifiers, (or op-amps). This is because this type of amplifier is used in circuits that perform mathematical algorithmic functions, or "operations" on input signals to obtain specific types of output signals. A typical op-amp has differential inputs (one "inverting", one, "non-inverting" relative to the output) and one output. An idealised op-amp has the following characteristics:
  - Infinite input impedance (so as to not load circuitry it is sampling as a control input)
  - Zero output impedance
  - Infinite gain
  - Zero propagation delay

The performance of an op-amp with these characteristics would be entirely defined by the (usually passive) components forming a negative feedback loop around it, that is, *the amplifier itself has no effect on the output*.

Today, op-amps are usually provided as integrated circuits, rather than constructed from discrete components. All real-world op-amps fall short of the idealised specification above – but some modern components have remarkable performance and come close in some respects.

Interstage Coupling Method

Amplifiers are sometimes classified by the coupling method of the signal at the input, output, or between stages. Different types of these include:

Resistive-capacitive (RC) coupled amplifier, using a network of resistors and capacitors

By design these amplifiers cannot amplify DC signals as the capacitors block the DC component of the input signal. RC-coupled amplifiers were used very often in circuits with vacuum tubes or discrete transistors. In the days of the integrated circuit a few more transistors on a chip are much cheaper and smaller than a capacitor.

Inductive-capacitive (LC) coupled amplifier, using a network of inductors and capacitors

This kind of amplifier is most often used in selective radio-frequency circuits.

Transformer coupled amplifier, using a transformer to match impedances or to decouple parts of the circuits

Quite often LC-coupled and transformer-coupled amplifiers cannot be distinguished as a transformer is some kind of inductor.

Direct coupled amplifier, using no impedance and bias matching components

This class of amplifier was very uncommon in the vacuum tube days when the anode (output) voltage was at greater than several 100 V and the grid (input) voltage at a few volts minus. So they were only used if the gain was specified down to DC (e.g., in an oscilloscope). In the context of modern electronics developers are encouraged to use direcly coupled amplifiers whenever possible.

Frequency Range

- Depending on the frequency range and other properties amplifiers are designed according to different principles.
- Frequency ranges down to DC are only used when this property is needed. DC amplification leads to specific complications that are avoided if possible.
- Depending on the frequency range specified different design principles must be used. Up to the MHz range only "discrete" properties need be considered; e.g., a terminal has an input impedance.

- As soon as any connection within the circuit gets longer than perhaps 1% of the wavelength of the highest specified frequency (e.g., at 100 MHz the wavelength is 3 m, so the critical connection length is approx. 3 cm) design properties radically change. For example, a specified length and width of a PCB trace can be used as a selective or impedance-matching entity.
- Above a few 100 MHz, it gets difficult to use discrete elements, especially inductors. In most cases PCB traces of very closely defined shapes are used instead.

Type of Load

- Untuned
  - Audio
  - Video

Tuned (RF amps) - used for amplifying a single radio frequency or band of frequencies

Implementation

Amplifiers are implemented using active elements of different kinds:

- The first active elements were relays. They were for example used in trans-continental telegraph lines: A weak current was used to switch the voltage of a battery to the outgoing line.
- For transmitting audio, carbon microphones were used as the active element. This was used to modulate a radio-frequency source in one of the first AM audio transmissions, by Reginald Fessenden on Dec. 24, 1906.
- Up to the early 1970s, most amplifiers used vacuum tubes (*valves* in the UK). Today, tubes are only generally used for very high-power, high-frequency amplifiers and for specialist audio applications, in which field they have recently achieved a new popularity. Many broadcast transmitters still use vacuum tubes. Additionally, their imperviousness to electromagnetic flash damage may have led to their retention in certain defence contexts.
- In the 1960s, the transistor started to take over. These days, discrete transistors are still used in high-power amplifiers and in specialist audio devices.
- Beginning in the 1970s, more and more transistors were connected on a single chip therefore creating the integrated circuit. Nearly all amplifiers commercially available today are based on integrated circuits.

For exotic purposes, other active elements have been used. For example, in the early days of the communication satellite parametric amplifiers were used. The core circuit was a diode whose capacity was changed by an RF signal created locally. Under certain conditions, this RF signal provided energy that was modulated by the extremely weak satellite signal received at the earth station. The operating principle of a parametric amplifier is somewhat similar to the principle by which children keep their swings in motion: as long as the swing moves you only need to change a parameter of the swinging entity; e.g., you must move your centre of gravity up and down. In our case, the capacity of the diode is changed periodically.

Power amplifier classes

Angle of flow or conduction angle

Power amplifier circuits (output stages) are classified as A, B, AB and C for analog designs, and class D and E for switching designs based upon the conduction angle or *angle of flow*, Θ, of the input signal through the amplifying device, that is, the portion of the input signal cycle during which the amplifying device conducts. The image of the conduction angle is derived from amplifying a sinusoidal signal. (If the device is always on, Θ = 360°.) The angle of flow is closely related to the amplifier power efficiency. The various classes are introduced below, followed by more detailed discussion under individual headings later on.

Class A

100% of the input signal is used (conduction angle Θ = 360° or 2π; i.e., the active element works in its linear range all of the time). Where efficiency is not a consideration, most small signal linear amplifiers are designed as Class A, which means that the output devices are always in the conduction region. Class A amplifiers are typically more linear and less complex than other types, but are very inefficient. This type of amplifier is most commonly used in small-signal stages or for low-power applications (such as driving headphones).

Class B

50% of the input signal is used (Θ = 180° or π; i.e., the active element works in its linear range half of the time and is more or less turned off for the other half). In most Class B, there are two output devices (or sets of output devices), each of which conducts alternately (push–pull) for exactly 180° (or half cycle) of the input signal; selective RF amplifiers can also be implemented using a single active element.

These amplifiers are subject to *crossover distortion* if the transition from one active element to the other is not perfect, as when two complementary transistors (i.e., one PNP, one NPN) are connected as two emitter followers with their base and emitter terminals in common, requiring the base voltage to slew across the region where both devices are turned off.

Class AB

Here the two active elements conduct more than half of the time as a means to reduce the cross-over distortions of Class B amplifiers. In the example of the complementary emitter followers a bias network allows for more or less quiescent current thus providing an operating point somewhere between Class A and Class B. Sometimes a figure is added (e.g., AB_1 or AB_2) with higher figures implying a higher quiescent current and therefore more of the properties of Class A.

Class D

Main article: Switching amplifier

These use switching to achieve a very high power efficiency (more than 90% in modern designs). By allowing each output device to be either fully on or off, losses are minimized. The analog output is created by pulse-width modulation; i.e., the active element is switched on for shorter or longer intervals instead of modifying its resistor. There are more complicated switching schemes like sigma-delta modulation, to improve some performance aspects like lower distortions or better efficiency.

Other classes

There are several other amplifier classes, although they are mainly variations of the previous classes. For example, Class G and Class H amplifiers are marked by variation of the supply rails (in discrete steps or in a continuous fashion, respectively) following the input signal. Wasted heat on the output devices can be reduced as excess voltage is kept to a minimum. The amplifier that is fed with these rails itself can be of any class. These kinds of amplifiers are more complex, and are mainly used for specialized applications, such as very high-power units. Also, Class E and Class F amplifiers are commonly described in literature for radio frequencies applications where efficiency of the traditional classes deviate substantially from their ideal values. These classes use harmonic tuning of their output networks to achieve higher efficiency and can be considered a subset of Class C due to their conduction angle characteristics.

More detail on the various classes is provided below.

Class A

Class A amplifying devices operate over the whole of the input cycle such that the output signal is an exact scaled-up replica of the input with no clipping. Class A amplifiers are the usual means of implementing small-signal amplifiers. They are not very efficient; a theoretical maximum of 50% is obtainable with inductive output coupling and only 25% with capacitive coupling.

In a Class A circuit, the amplifying element is biased so the device is always conducting to some extent, and is operated over the most linear portion of its characteristic curve (known as its transfer characteristic or transconductance curve). Because the device is always conducting, even if there is no input at all, power is drawn from the power supply. This is the chief reason for its inefficiency.

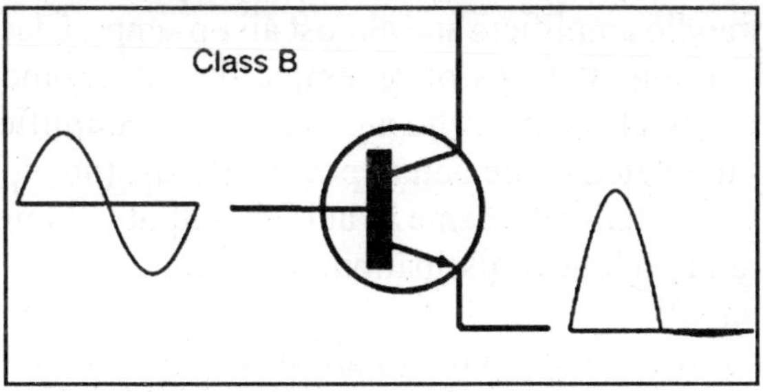

Fig.. Class A Amplifier

If high output powers are needed from a Class A circuit, the power waste (and the accompanying heat) will become significant. For every watt delivered to the load, the amplifier itself will, at best, dissipate another watt. For large powers this means very large and expensive power supplies and heat sinking. Class A designs have largely been superseded for audio amplifiers, though some audiophiles believe that Class A gives the best sound quality, due to it being operated in as linear a manner as possible which provides a small market for expensive high fidelity Class A amps. In addition, some aficionados prefer thermionic valve (or "tube") designs instead of transistors, for several claimed reasons:

- Tubes are more commonly used in class A designs, which have an asymmetrical transfer function. This means that distortion of a sine wave creates both odd- and even-numbered harmonics. The claim is that this sounds more "musical" than the higher level of odd harmonics produced by a symmetrical push–pull amplifier.
- Though good amplifier design can reduce harmonic distortion patterns to almost nothing, distortion is essential to the sound of electric guitar amplifiers, for example, and is held by recording engineers to offer more flattering microphones and to enhance "clinical-sounding" digital technology.
- Valves use many more electrons at once than a transistor, and so statistical effects lead to a "smoother" approximation of the true

waveform. Junction field-effect transistors (JFETs) have similar characteristics to valves, so these are found more often in high quality amplifiers than bipolar transistors. Historically, valve amplifiers often used a Class A power amplifier simply because valves are large and expensive; many Class A designs use only a single device.

Transistors are much cheaper, and so more elaborate designs that give greater efficiency but use more parts are still cost-effective. A classic application for a pair of class A devices is the long-tailed pair, which is exceptionally linear, and forms the basis of many more complex circuits, including many audio amplifiers and almost all op-amps. Class A amplifiers are often used in output stages of op-amps; they are sometimes used as medium-power, low-efficiency, and high-cost audio amplifiers. The power consumption is unrelated to the output power. At idle (no input), the power consumption is essentially the same as at high output volume. The result is low efficiency and high heat dissipation.

Class B and AB

Class B amplifiers only amplify half of the input wave cycle. As such they create a large amount of distortion, but their efficiency is greatly improved and is much better than Class A. Class B has a maximum theoretical efficiency of 78.5% (i.e., $\pi/4$). This is because the amplifying element is switched off altogether half of the time, and so cannot dissipate power. A single Class B element is rarely found in practice, though it can be used in RF power amplifier where the distortion levels are less important. However Class C is more commonly used for this.

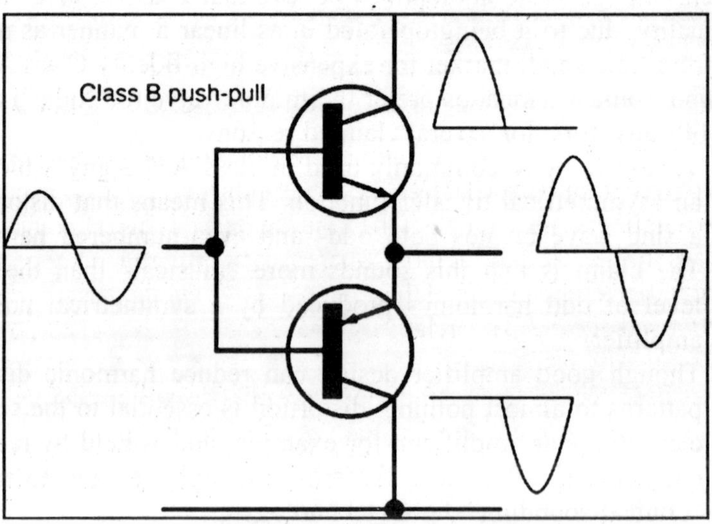

Fig. Class B Amplifier

A practical circuit using Class B elements is the complementary pair or "push–pull" arrangement. Here, complementary or quasi-complementary devices are used to each amplify the opposite halves of the input signal, which is then recombined at the output. This arrangement gives excellent efficiency, but can suffer from the drawback that there is a small mismatch at the "joins" between the two halves of the signal. This is called crossover distortion. An improvement is to bias the devices so they are not completely off when they're not in use. This approach is called Class AB operation.

In Class AB operation, each device operates the same way as in Class B over half the waveform, but also conducts a small amount on the other half. As a result, the region where both devices simultaneously are nearly off (the "dead zone") is reduced. The result is that when the waveforms from the two devices are combined, the crossover is greatly minimised or eliminated altogether.

Class AB sacrifices some efficiency over class B in favor of linearity, so will always be less efficient (below 78.5%). It is typically much more efficient than class A.

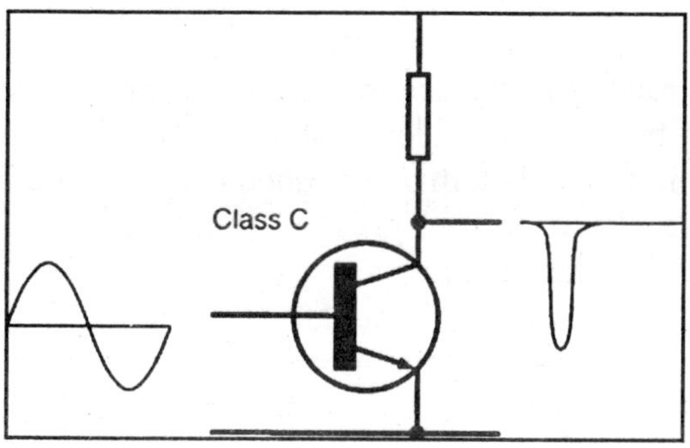

Fig. Class B push–pull Amplifier

Class B or AB push–pull circuits are the most common design type found in audio power amplifiers. Class AB is widely considered a good compromise for audio amplifiers, since much of the time the music is quiet enough that the signal stays in the "class A" region, where it is amplified with good fidelity, and by definition if passing out of this region, is large enough that the distortion products typical of class B are relatively small. The crossover distortion can be reduced further by using negative feedback. Class B and AB amplifiers are sometimes used for RF linear amplifiers as well. Class B amplifiers are also favored in battery-operated devices, such as transistor radios.

Digital Class B

A limited power output Class-B amplifier with a single-ended supply rail of 5±0.5 V.

Class C

Class C amplifiers conduct less than 50% of the input signal and the distortion at the output is high, but high efficiencies (up to 90%) are possible. Some applications (for example, megaphones) can tolerate the distortion. A much more common application for Class C amplifiers is in RF transmitters, where the distortion can be vastly reduced by using tuned loads on the amplifier stage. The input signal is used to roughly switch the amplifying device on and off, which causes pulses of current to flow through a tuned circuit.

The Class C amp. has two modes of operation: tuned, and untuned. The diagram below shows a waveform from a simple class C circuit without the tuned load. This is called untuned operation, and the analysis of the waveforms shows the massive distortion that appears in the signal. When the proper load (e.g., a pure inductive-capacitive filter) is used, two things happen. The first is that the output's bias level is clamped, so that the output variation is centered at one-half of the supply voltage. This is why tuned operation is sometimes called a *clamper*. This action of elevating bias level allows the waveform to be restored to its proper shape, allowing a complete waveform to be re-established despite having only a one-polarity supply. This is directly related to the second phenomenon: the waveform on the centre frequency becomes much less distorted. The distortion that is present is dependent upon the bandwidth of the tuned load, with the centre frequency seeing very little distortion, but greater attenuation the farther from the tuned frequency that the signal gets.

The tuned circuit will only resonate at particular frequencies, and so the unwanted frequencies are dramatically suppressed, and the wanted full signal (sine wave) will be extracted by the tuned load (e.g., a high-quality bell will ring at a particular frequency when it is hit periodically with a hammer). Provided the transmitter is not required to operate over a very wide band of frequencies, this arrangement works extremely well. Other residual harmonics can be removed using a filter.

Class C Amplifier

Class D

Block diagram of a basic switching or PWM (Class-D) amplifier.

Class D amplifiers are much more efficient than Class AB power amplifiers. As such, Class D amplifiers do not need large transformers and heavy heatsinks, which means that they are smaller and lighter in weight than an equivalent Class AB amplifier. All power devices in a Class D amplifier are operated in on/off mode. Output stages such as those used in pulse generators are examples of class D amplifiers. The term usually applies to devices intended to reproduce signals with a bandwidth well below the switching frequency.

These amplifiers use pulse width modulation, pulse density modulation (sometimes referred to as pulse frequency modulation) or more advanced form of modulation such as Delta-sigma modulation (for example, in the Analog Devices AD1990 Class-D audio power amplifier).

The input signal is converted to a sequence of pulses whose averaged value is directly proportional to the instantaneous amplitude of the signal. The frequency of the pulses is typically ten or more times the highest frequency of interest in the input signal. The output of such an amplifier contains unwanted spectral components (that is, the pulse frequency and its harmonics) which must be removed by a passive filter. The resulting filtered signal is then an amplified replica of the input.

The main advantage of a class D amplifier is power efficiency. Because the output pulses have a fixed amplitude, the switching elements (usually MOSFETs, but valves and bipolar transistors were once used) are switched either on or off, rather than operated in linear mode. This means that very little power is dissipated by the transistors, except during the very short interval between the on and off states. The wasted power is low because the instantaneous power dissipated in the transistor is the product of voltage and current, and one or the other is almost always close to zero. The lower losses permit the use of a smaller heat sink while the power supply requirements are lessened too.

Class D amplifiers can be controlled by either analog or digital circuits. The digital control introduces additional distortion called *quantization error* caused by its conversion of the input signal to a digital value.

Class D amplifiers have been widely used to control motors, and almost exclusively for small DC motors, but they are now also used as audio amplifiers, with some extra circuitry to allow analogue to be converted to a much higher frequency pulse width modulated signal. The relative difficulty of achieving good audio quality means that nearly all are used in applications where quality is not a factor, such as modestly-priced bookshelf audio systems and "DVD-receivers" in mid-price home theater systems.

High quality Class D audio amplifiers are now, however, starting to appear in the market:

- Tripath have called their revised Class D designs Class T.
- Bang and Olufsen's ICEPower Class D system has been used in the Alpine PDX range and some Pioneer's PRS range and for other manufacturers' equipment.

These revised designs have been said to rival good traditional AB amplifiers in terms of quality.

Before these higher quality designs existed an earlier use of Class D amplifiers and prolific area of application was high-powered, subwoofer amplifiers in cars. Because subwoofers are generally limited to a bandwidth of no higher than 150 Hz, the switching speed for the amplifier does not have to be as high as for a full range amplifier. The drawback with Class D designs being used to power subwoofers is that their output filters (typically inductors that convert the pulse width signal back into an analogue waveform) lower the damping factor of the amplifier.

This means that the amplifier cannot prevent the subwoofer's reactive nature from lessening the impact of low bass sounds (as explained in the feedback part of the Class AB section). Class D amplifiers for driving subwoofers are relatively inexpensive, in comparison to Class AB amplifiers. A 1000 watt Class D subwoofer amplifier that can operate at about 80% to 95% efficiency costs about $250 USD, much less than a Class AB amplifier of this power, which would cost several thousand dollars.

The letter *D* used to designate this amplifier class is simply the next letter after *C*, and does not stand for *digital*. Class D and Class E amplifiers are sometimes mistakenly described as "digital" because the output waveform superficially resembles a pulse-train of digital symbols, but a Class D amplifier merely converts an input waveform into a continuously pulse-width modulated (square wave) analog signal. (A digital waveform would be pulse-code modulated.)

Special classes

Class E

The class E/F amplifier is a highly efficient switching power amplifier, typically used at such high frequencies that the switching time becomes comparable to the duty time. As said in the class-D amplifier the transistor is connected via a serial-LC-circuit to the load, and connected via a large L (inductance) to the supply voltage. The supply voltage is connected to ground via a large capacitor to prevent any RF-signals leaking into the supply. The class-E amplifier adds a *C* between the transistor and ground and uses a defined L_1 to connect to the supply voltage.

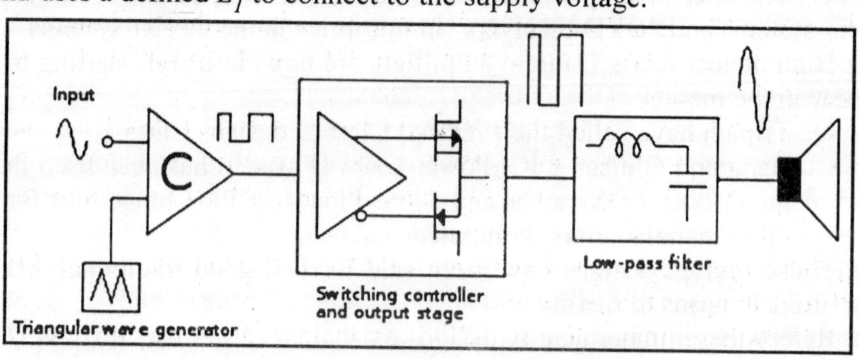

Class E Amplifier

The following description ignores DC, which can be added afterwards easily. The above mentioned *C* and L are in effect a parallel LC-circuit to ground. When the transistor is on, it pushes through the serial LC-circuit into the load and some current begins to flow to the parallel LC-circuit to ground. Then the serial LC-circuit swings back and compensates the current into the parallel LC-circuit. At this point the current through the transistor is zero and it is switched off. Both LC-circuits are now filled with energy in the C and the L_0. The whole circuit performs a damped oscillation. The damping by the load has been adjusted so that some time later the energy from the Ls is gone into the load, but the energy in both C_0 peaks at the original value, to in turn restore the original voltage, so that the voltage across the transistor is zero again and it can be switched on.

With load, frequency, and duty cycle (0.5) as given parameters and the constraint that the voltage is not only restored, but peaks at the original voltage, the four parameters (L, L_0, C and C_0) are determined. The class F-amplifier takes the finite on resistance into account and tries to make the current touch the bottom at zero. This means the voltage and the current at the transistor are symmetric with respect to time. The Fourier transform allows an elegant formulation to generate the complicated LC-networks. It says that the first harmonic is passed into the load, all even harmonics are shorted and all higher odd harmonics are open.

Class F and the even harmonics

In push–pull amplifiers and in CMOS, the even harmonics of both transistors just cancel. Experiment shows that a square wave can be generated by those amplifiers and theory shows that square waves do consist of odd harmonics only. In a class D amplifier, the output filter blocks all harmonics; i.e., the harmonics see an open load. So even small currents in the harmonics suffice to generate a voltage square wave. The current is in phase with the voltage applied to filter, but the voltage across the transistors is out of phase. Therefore, there is a minimal overlap between current through the transistors and voltage across the transistors. The sharper the edges the lower the overlap.

While class D sees the transistors and the load as two separate modules, the class F admits imperfections like the parasitics of the transistor and tries to optimise the global system to have a high impedance at the harmonics. Of course there has to be a finite voltage across the transistor to push the current across the on state resistance. Because the combined current through both transistors is mostly in the first harmonic it looks like a sine. That means that in the middle of the square the maximum of current has to flow, so it may make sense to have a dip in the square or in other words to allow some over swing of the voltage square wave. A class F load network by definition has to transmit below a cut off frequency and to reflect above.

Any frequency lying below the cut off and having its second harmonic above the cut off can be amplified, that is an octave bandwidth. On the other hand, an inductive-capacitive series circuit with a large inductance and a tunable capacitance may be simpler to implement. By reducing the duty cycle below 0.5, the output amplitude can be modulated. The voltage square waveform will degrade, but any overheating is compensated by the lower overall power flowing. Any load mismatch behind the filter can only act on the first harmonic current waveform, clearly only a purely resistive load makes sense, then the lower the resistance the higher the current.

Class F can be driven by sine or by a square wave, for a sine the input can be tuned by an inductor to increase gain. If class F is implemented with a single transistor, the filter is complicated to short the even harmonics. All previous designs use sharp edges to minimise the overlap.

Class E uses a significant amount of second harmonic voltage. The second harmonic can be used to reduce the overlap with edges with finite sharpness. For this to work energy on the second harmonic has to flow from the load into the transistor, and no source for this is visible in the circuit diagram. In reality, the impedance is mostly reactive and the only reason for it is that class E is a class F amplifier with a much simplified load network and thus has to deal with imperfections.

In many amateur simulations of class E amplifiers, sharp current edges are assumed nullifying the very motivation for class E and measurements near the transit frequency of the transistors show very symmetric curves, which look much similar to class F simulations. The class E amplifier was invented in 1972 by Nathan O. Sokal and Alan D. Sokal, and details were first published in 1975. Some earlier reports on this operating class have been published in Russian.

Class G and H

There is a variety of amplifier designs that couple a class AB output stage with other more efficient techniques to achieve a higher efficiency with low distortion. These designs are common in large audio amplifiers since the heatsinks and power transformers would be prohibitively large (and costly) without the increase in efficiency. The terms "class G" and "class H" are used interchangeably to refer to different designs, varying in definition from one manufacturer or paper to another.

Class G amplifiers (which use "rail switching" to decrease power consumption and increase efficiency) are more efficient than class AB amplifiers. The class G amplifier has several power rails at different voltages, and switches between rails as the signal output approaches each. Thus the amplifier increases efficiency by reducing the wasted power at the output transistors.

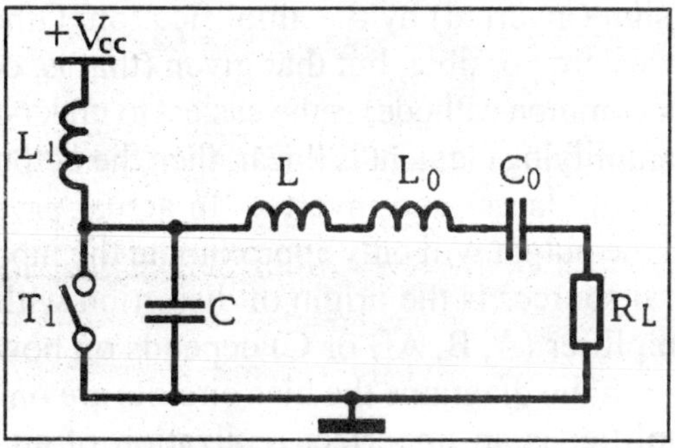

A Class H amplifier takes the idea of Class G one step further creating an infinitely variable supply rail. This is done by modulating the supply rails so that the rails are only a few volts larger than the output signal at any given time.

The output stage operates at its maximum efficiency all the time. Switched mode power supplies can be used to create the tracking rails. Significant efficiency gains can be achieved but with the drawback of more complicated supply design and reduced THD performance.

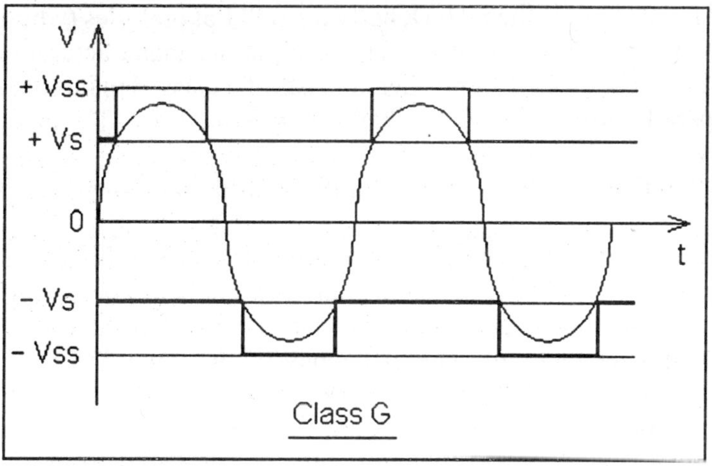

Efficiency Class H

The classes can be most easily understood using the diagrams in each section below. For the sake of illustration, a bipolar junction transistor is shown as the amplifying device, but in practice this could be a MOSFET or vacuum tube device. In an analog amplifier (the most common kind), the signal is applied to the input terminal of the device (base, gate or grid), and this causes a proportional output drive current to flow out of the output terminal. The output drive current comes from the power supply.

The voltage signal shown is thus a larger version of the input, but has been changed in sign (inverted) by the amplification. Other arrangements of amplifying device are possible, but that given (that is, common emitter, common source or common cathode) is the easiest to understand and employ in practice. If the amplifying element is linear, then the output will be faithful copy of the input, only larger and inverted. In actual practice, transistors are not linear, and the output will only approximate the input. Non-linearity from any of several sources is the origin of distortion within an amplifier. Which class of amplifier (A, B, AB or C) depends on how the amplifying device is biased — in the diagrams the bias circuits are omitted for clarity.

Any real amplifier is an imperfect realization of an ideal amplifier. One important limitation of a real amplifier is that the output it can generate is ultimately limited by the power available from the power supply. An amplifier will saturate and clip the output if the input signal becomes too large for the amplifier to reproduce or if operational limits for a device are exceeded.

Doherty Amplifiers

A hybrid configuration receiving new attention is the Doherty amplifier, invented in 1934 by William H. Doherty for Bell Laboratories (whose sister company, Western Electric, was then an important manufacturer of radio transmitters). The Doherty amplifier consists of a class-B main (or *carrier*) stage in parallel with a class-C auxiliary (or *peaking*) stage. The input signal is split evenly to drive the two amplifiers, and a combining network sums the two output signals and corrects for phase differences between the two amplifiers.

During periods of low signal level, the class-B amplifier efficiently operates on the signal and the class-C amplifier is inactive and consumes no power. During high signal peaks the class-B amplifier saturates and the class-C amplifier kicks in. The efficiency of previous AM transmitter designs was proportional to modulation, but with average modulation typically 20 per cent, transmitters were limited to less than 50 per cent efficiency. In Doherty's design, even with zero modulation a transmitter could achieve at least 60 per cent efficiency.

As a successor to Western Electric for broadcast transmitters, the Doherty concept was considerably refined by Continental Electronics Manufacturing Company of Dallas, Texas. Perhaps the ultimate refinement was the screen grid modulation scheme invented by Joseph B. Sainton. The Sainton amplifier consists of a Class C main (or *carrier*) stage in parallel with a Class C auxiliary (or *peak*) stage. The stages are split and combined through 90 degree phase shifting networks as in the Doherty amplifier. The unmodulated radio frequency carrier is applied to the control grids of both tubes.

Carrier modulation is applied to the screen grids of both tubes. The bias of the carrier and peak tubes are different, and are established so that the peak tube is quiescent when modulation is absent (and the amplifier is producing rated unmodulated carrier power) whereas both tubes contribute twice the rated carrier power during 100 per cent modulation (as four times the carrier power is required to achieve 100 per cent modulation). As both tubes operate in Class C, a significant improvement in efficiency is thereby achieved in the final stage. And, as the tetrode carrier and peak tubes require very little drive power, a significant improvement in efficiency within the driver stage is achieved as well (317C, et al.). The released version of the Sainton amplifier employs a cathode follower modulator, not a push-pull modulator. Previous Continental Electronics designs, by James O. Weldon and others, retained most of the characteristics of the Doherty amplifier but added screen grid modulation of the driver (317B, et al.).

The Doherty amplifier remains in use in very-high-power AM transmitters, but for lower-power AM transmitters, vacuum-tube amplifiers in general were eclipsed in the 1980s by arrays of solid-state amplifiers, which could be switched on and off with much finer granularity in response to the requirements of the input audio. However, interest in the Doherty configuration has been revived by cellular-telephone and wireless-Internet applications where the sum of several constant-envelope users creates an aggregate AM result. The main challenge of the Doherty amplifier for digital transmission modes is in aligning the two stages and getting the class-C amplifier to turn on and off very quickly.

Recently, Doherty amplifiers have found widespread use in cellular base station transmitters for GHz frequencies. Implementations for transmitters in mobile devices have also been demonstrated.

Other Classes

Several audio amplifier manufacturers have started "inventing" new classes as a way to differentiate themselves. These class names usually do not reflect any revolutionary amplification technique, and are used mostly for marketing purposes. This can easily be determined by the fact that the class name is trademarked or copyrighted. For example, Crown's K and I-Tech Series as well as several other models utilise Crown's patented Class-I (or BCA) technology. Lab.gruppen use a form of class D amplifier called class TD or Tracked Class D which tracks the waveform to more accurately amplify it without the drawbacks of traditional class D amplifiers.

"Class T" is a trademark of TriPath company, which manufactures audio amplifier ICs. This new class "T" is a revision of the common class D amplifier, but with changes to ensure fidelity over the full audio spectrum, unlike traditional class D designs. It operates at different frequencies depending on the power output, with values ranging from as low as 200 kHz to 1.2 MHz, using a proprietary modulator.

"Class Z" is a trademark of Zetex semiconductor and is a direct-digital-feedback technology.

Amplifier Circuit

The practical amplifier circuit below could be the basis for a moderate-power audio amplifier. It features a typical (though substantially simplified) design as found in modern amplifiers, with a class AB push–pull output stage, and uses some overall negative feedback. Bipolar transistors are shown, but this design would also be realizable with FETs or valves.

A Practical Amplifier Circuit

The input signal is coupled through capacitor C1 to the base of transistor Q1. The capacitor allows the AC signal to pass, but blocks the DC bias voltage established by resistors R1 and R2 so that any preceding circuit is not affected by it. Q1 and Q2 form a differential amplifier (an amplifier that multiplies the difference between two inputs by some constant), in an arrangement known as a long-tailed pair. This arrangement is used to conveniently allow the use of negative feedback, which is fed from the output to Q2 via R7 and R8.

The negative feedback into the difference amplifier allows the amplifier to compare the input to the actual output. The amplified signal from Q1 is directly fed to the second stage, Q3, which is a common emitter stage that provides further amplification of the signal and the DC bias for the output stages, Q4 and Q5. R6 provides the load for Q3 (A better design would probably use some form of active load here, such as a constant-current sink).

So far, all of the amplifier is operating in Class A. The output pair are arranged in Class AB push–pull, also called a complementary pair. They provide the majority of the current amplification and directly drive the load, connected via DC-blocking capacitor C2. The diodes D1 and D2 provide a small amount of constant voltage bias for the output pair, just biasing them into the conducting state so that crossover distortion is minimized. That is, the diodes push the output stage firmly into class-AB mode (assuming that the base-emitter drop of the output transistors is reduced by heat dissipation).

This design is simple, but a good basis for a practical design because it automatically stabilises its operating point, since feedback internally operates from DC up through the audio range and beyond. Further circuit elements would probably be found in a real design that would roll off the frequency response above the needed range to prevent the possibility of unwanted oscillation. Also, the use of fixed diode bias as shown here can cause problems if the diodes are not both electrically and thermally matched to the output transistors — if the output transistors turn on too much, they can easily overheat and destroy themselves, as the full current from the power supply is not limited at this stage.

A common solution to help stabilise the output devices is to include some emitter resistors, typically an ohm or so. Calculating the values of the circuit's resistors and capacitors is done based on the components employed and the intended use of the amp.

Notes on implementation

Real world amplifiers are imperfect.

- One consequence is that the power supply itself may influence the output, and must itself be considered when designing the amplifier
- The amplifier circuit has an "open loop" performance, that can be described by various parameters (gain, slew rate, output impedance, distortion, bandwidth, signal to noise ratio...)
- Many modern amplifiers use negative feedback techniques to hold the gain at the desired value.

Different methods of supplying power result in many different methods of bias. Bias is a technique by which the active devices are set up to operate in a particular regime, or by which the DC component of the output signal is set to the midpoint between the maximum voltages available from the power supply. Most amplifiers use several devices at each stage; there are typically matched in specifications except for polarity. Matched inverted polarity devices are called complementary pairs. Class A amplifiers generally use only one device, unless the power supply is set to provide both positive and negative voltages, in which case a dual device symmetrical design may be used. Class C amps, by definition, use a single polarity supply.

Amplifiers often have multiple stages in cascade to increase gain. Each stage of these designs may be a different type of amp to suit the needs of that stage. For instance, the first stage might be a Class A stage, feeding a class AB push–pull second stage, which then drives a class G final output stage, taking advantage of the strengths of each type, while minimizing their weaknesses.

Power Amplifier

The term "power amplifier" is a relative term with respect to the amount of power delivered to the load and/or sourced by the supply circuit. In general a power amplifier is designated as the last amplifier in a transmission chain (the *output stage*) and is the amplifier stage that typically requires most attention to power efficiency.

Vacuum Tube (valve) Amplifiers

According to Symons, while semiconductor amplifiers have largely displaced valve amplifiers for low power applications, valve amplifiers are much more cost effective in high power applications such as "radar, countermeasures equipment, or communications equipment". Many microwave amplifiers are specially designed valves, such as the klystron, gyrotron, traveling wave tube, and crossed-field amplifier, and these microwave valves provide much greater single-device power output at microwave frequencies than solid-state devices.

A valve amplifier or tube amplifier is a type of electronic amplifier that makes use of vacuum tubes to increase the power and/or amplitude of a signal. They are typically (but not exclusively) used for sound amplification, either in home stereo hi-fi amplifiers or for electric guitar amps, for radio frequency signals.

Low to medium power valve amplifiers for frequencies below the microwaves were largely replaced by solid state amplifiers during the 1960s and 1970s. Valve amplifiers are used for applications such as guitar amplifiers, satellite transponders such as DirecTV and GPS systems, audiophile stereo amplifiers, military applications (such as target acquisition and radar) and very high power radio and UHF television transmitters.

Origins

Until the invention of the transistor in 1947, all practical amplifiers were made using thermionic valves. The simplest valve was invented by John Ambrose Fleming while working for the Marconi Company in London in 1904 and named the Diode, as it had two electrodes. The diode conducted electricity in one direction only and was used as a radio detector and a rectifier.

Although he may not have at first realised the significance of his invention it was Lee De Forest who added a third electrode and invented the first electronic amplifying device, the Triode which he named the 'Audion'. This additional 'control grid' modulates the current that flows between cathode and anode for a given voltage between the cathode and anode. The relationship between current flow and plate and grid voltage is often represented as a series of "characteristic curves" on a diagram. Depending on the other components in the circuit this modulated current flow can be used to provide current or voltage gain.

The first application of valve amplification was in the regeneration of long distance telephony signals. Later, valve amplification was applied to the 'wireless' market that began in the early thirties. In due course amplifiers for music and later television were also built exclusively using valves until the 1950s. During this period power levels were usually very low (a few watts) and radios often used headphones only, with no loudspeaker at all.

The overwhelmingly dominant circuit topology during this period was the single-ended triode gain stage, operating in class A, which gave very good sound (and reasonable measured distortion performance) despite extremely simple circuitry with very few components: important at a time when components were hand made and extremely expensive. Prior to World War II almost all valve amplifiers were of low gain and with linearity dependent entirely on the inherent linearity of the tube itself, typically 5% distortion at full power.

Post-War Developments

Increasing post-war affluence and industrialised production economies, dramatic technical progress stimulated by the war, and for the first time a substantial and expanding consumer market brought more advanced valve designs to market at affordable prices, with the result that the 1960's saw the increasing spread of Gramophone players, and ultimately the beginnings of "HiFi" able to drive full frequency range loudspeakers (for the first time often with multiple drivers for different frequencies) to significant volume levels, combined with the spread of B&W TV, to produce a 'golden age' in Valve development and also in the development of Valve amplifier circuits.

Negative feedback (NFB) was invented by Harold Stephen Black in 1927, but initially little used since at that time gain was at a premium. This technique allows amplifiers to trade gain for reduced distortion levels (and also gave other benefits such as reduced output impedance). The introduction of the Williamson amplifier in 1947, which was extremely advanced in many respects including very successful use of NFB, was a turning point in audio power amplifier design, operating a push pull output circuit in class AB1 to give performance far ahead of its time.

Similar topologies with only minor variations (notably different phase splitter arrangements and the "ultra-Linear" transformer connection for tetrodes) rapidly became widespread and this family of designs remains the dominant high power amplifier topology to this day for music application.This period also saw continued growth in civilian radio, with valves being used for both transmitters and receivers.

Decline

From the 70s the silicon transistor became increasingly pervasive and valve production was sharply ramped down, with the notable exception of

Cathode Ray Tubes, and a dramatically rationalized range of valves for amplifier applications, low power tubes being mostly dual triodes (ECCnn, 12Ax7 series) plus the EF86 pentode, power tubes mostly being Beam Tetrode and Pentodes (EL84, EL34, KT88/6550, 6L6), in both cases with indirect heating and this rationalized set of types remains the core of all subsequent production and remain in production (in the east) today.

The Soviets retained valves to a much greater extent than the west during the cold war, for the majority of their communications and military amplification requirements, in part due to tubes' ability to withstand instantaneous overloads (notably due to a nuclear detonation) that would destroy a transistor.

The dramatic reduction in size, power consumption, reduced distortion levels and above all cost of electronics products based on Transistors has made valves obsolete for mainstream products since ~ the 1970s, although valves remained in niche (mainly high power RF transmitters) applications for somewhat longer. However, the difficulty in producing transistors with good gain and efficiency at very high frequencies, combined with the fragility of transistors (problems such as thermal runaway), resulted in tubes being retained for longer in high power and high frequency applications, notably large radio (and TV) transmitters and guitar amplifier for many more years. However, developments in semiconductor production have now mostly closed that market also.

Resurgence

In audio applications, valves continue to be highly desired by some users, both in the higher-end home audio market and in the guitar amplifier market. Amongst stereo enthusiasts, there is a subgroup of audio buffs who advocate the use of tube amplifiers for home listening; they argue that tube amplifiers produce a "warmer" or more "natural" sound. Companies in Russia, China and Eastern Europe continue to produce valves to cater to this market.

In the guitar amplifier market, some performers continue to use tube amps in the 2000s, including in folk, blues, roots rock, and in harder genres such as metal, where Marshall tube amps are used to create heavy distortion. Audio engineers suggest that the subjectively-pleasing aspects of tube amplification may be due to the non-linear overdrive that is produced with tubes. Tube amps have the following advantages over solid-state amps.

Compared to semiconductors, tubes have a very low "drift" (of specs) over a wide range of operating conditions, specifically high heat/high power. Semiconductors are very heat-sensitive by comparison and this fact usually leads to compromises in solid-state amplifier designs. When a tube fails, it is replaceable. While solid state devices are also replaceable, it is usually a much more involved process (i.e., having the amplifier tested by a professional, removing the faulty component, and replacing it).

For working musicians this is usually a huge problem by comparison to looking in the back of a tube amp at the tubes and simply replacing the faulty tube. In addition, tubes can easily be removed and tested, while transistors cannot. Tube amplifiers respond differently from transistor amplifiers when signal levels approach and reach the point of clipping. In a tube-powered amplifier, the transition from linear amplification to limiting is less abrupt than in a solid state unit, resulting in a less grating form of distortion at the onset of clipping. For this reason, some guitarists prefer the sound of an all-tube amplifier; the aesthetic properties of tube versus solid state amps, though, are a topic of debate in the guitarist community.

Characteristics

Valves are high voltage/low current devices in comparison with transistors (and especially MOSFETs). The high working voltage makes them well suited for radio transmitters, for example, and valves remain in use today for very high power radio transmitters, where there is still no other technology available. However, for most applications requiring an appreciable output current, a matching transformer is required. The transformer is a critical component and heavily influences the performance (and cost) of the amplifier.

Many power valves have good linearity but modest gain or transconductance. Signal amplifiers using tubes are capable of very high frequency response ranges - up to radio frequency. Indeed, many of the Directly Heated Single Ended Triode (DH-SET) audio amplifiers are in fact radio transmitting tubes designed to operate in the megahertz range. In practice, however, tube amplifier designs typically "couple" stages either capacitively, limiting bandwidth at the low end, or inductively with transformers, limiting the bandwidth at high end.

Advantages

- Very linear (especially triodes) making it viable to use them in low distortion linear circuits with little or no negative feedback.
- Extremely high input impedance (cf bipolar transistors but a characteristic shared by FETs).
- High voltage devices and thus inherently suitable for very high voltage circuits.
- Can be constructed on a scale that can dissipate large amounts of heat (some extreme devices even being water cooled). For this reason valves remained the only viable technology for very high power, and especially high power/high voltage applications

such as Radio & TV transmitters long into the age when transistors had displaced valves in most other applications. However, today these also are becoming obsolete.

- Electrically very robust, they can tolerate overloads for minutes which would destroy bipolar transistor systems in milliseconds. In the worst case a failed tube can simply be unplugged and replaced by the user.

Disadvantages

- Heater supplies are usually required for the cathodes, and dangerously high voltages are usually required for the anodes.
- They are significantly larger than equivalent solid-state transistors
- High impedance/low current output is unsuitable for direct drive of many real world loads, notably various forms of electric motors.
- Valved audio equipment is normally heavy because of the weight of transformers.
- Valves may have a shorter working life than solid state parts due to various failure mechanisms (cathode poisoning, breakages (i.e., open circuit) or shorts internally, notably of the heater or grid structures, or in the case of glass valves, physical breakage, although this should not be overstated; many valve types typically have operation lives in the tens of thousands of hours and an indefinite shelf life (many 60 year old tubes are still in regular use).
- Available in a single polarity only whereas transistors are available in complementary polarities (e.g., NPN/PNP), making possible many circuit configurations that cannot be realized directly with valves.
- In comparison to the lower impedance environment of transistors special consideration must be made to the physical layout of valve circuits in order to avoid instability and the introduction of noise from radio frequency interference and ac heater supplies.

Operation

All amplifier circuits are classified by "class of operation" as A, B, AB and C etc. However, the nature of valves results in significantly different circuit topologies and characteristics than transistor designs, and as a sweeping generalisation, valve amplifiers tend to operate in class A (or class AB1 with a heavy class A overlap), whereas transistor amplifiers tend to operate more in class B (there are however significant exceptions in both directions).

- The grid (where the input signal is presented) needs to be biased substantially negative with respect to the cathode (typically ~ -75 V). This makes it extremely difficult to direct-couple the output of one valve (typically sitting at about + 100 V) to the input of a following valve as is normally done in modern transistor designs. (cf transistors which are usually biased just a few volts positive), at a voltage between that of the collector and emitter, facilitating direct coupling)

- Valve stages thus need to be coupled using some component able to totally block and withstand several hundred volts, typically a capacitor, occasionally a coupling transformer, adding phase shifts and possibly coloration to the signal. These introduced phase shifts can become problematic in circuits that have feedback
- There is no valve analog of the complementary devices widely used in "totem pole" output stages of silicon circuits. This is because valves work based on the flow of electrons from the cathode to the anode and it is not possible to construct a hypothetical valve in which an "electron hole" migrated the other way. Push-pull valve topologies therefore typically require a transformer.
- The very high output impedance of valves (compared with transistors) usually demands the use of matching transformers if low impedance loads (notably loudspeakers or various forms of motor, such as cutting lathe heads, etc.) are to be driven. The transformer is used as the load, in place of the resistor usually used in small-signal and driver stages. NB the impedance of the transformer primary at the frequencies in use is much higher than the DC resistance of the windings, often kOhms. High performance transformers are however severe engineering compromises, are expensive, and in operation are far from ideal. Transformers dramatically increase the cost of a valve amplifier circuit compared to a direct-coupled transistor alternative.
- The (typically) much lower open loop gain but enhanced open loop linearity of valves, especially triodes, makes it possible to use little or no negative feedback in circuits whilst retaining acceptable or even excellent distortion performance (especially for small-signal circuits).

Topologies

- Linear small signal circuits almost invariably use a triode in the single ended gain stage topology (in class A), including the output stage (cf silicon circuits, notably the very widely use "op-amp" configuration, which normally have a totem pole output stage).

- Broadband valve amplifiers typically use class A1 or AB1,
- Modern high power output stages are usually push pull, often necessitating somc form of phase splitter to derive a differential/ balanced drive signal from a single ended input, typically followed by a further gain stage (the "driver") prior to the output tubes.
- Single ended" power stages using very large valves exist and dominate in radio transmitter applications. A sidebar is the observation that the niche "DH-SET" (directly heated single-ended triode) topology favored by some audiophiles are extremely simple and typically constructed using valve types originally designed for use in radio transmitters
- More complex topologies (notably the use of active loads) can improve linearity and frequency response (by removing miller capacitance effects).

Output Transformerless

Julius Futterman pioneered a type of amplifier known as "output transformerless" (OTL). These forego the typical output transformer by paralleling (electrically connecting and operating side-by-side) perhaps one dozen or more output tubes in an attempt to reduce effective plate resistance and more closely match it with speaker impedances (typically 8 ohms). This design and its various incarnations tend to require numerous tubes, run hot, and because they attempt to match impedances in a way fundamentally different from a transformer, they often have a unique sound quality.

Output Impedance

The high voltage/low current/High output impedance (Z out) of the output (anode circuit in the overwhelming majority of valve circuits) is suitable to drive another following valve stage, and can drive an antenna system that has been arranged to resonate at the required drive frequency, but usually is not suitable to drive low impedance loads, such as (notably) loudspeakers and motors. The main techniques are used to resolve this: the use of a matching transformer, and the use of negative feedback to reduce that active output impedance (in proportion to the amount of feedback applied). In combination these techniques can reduce the Z out from hundreds of ohms to a fraction of an ohm.

Applications

Audio Frequency (AF)/Broadband amplifiers

Valves remain in widespread use in guitar and high-end audio amplifiers due to the sound quality they produce, which is subjectively preferred by some users. They are largely obsolete for most other applications, mainly due to the cost effectiveness advantages of the transistor, and due to the lower weight and heat production of transistor amps.

Telephony

(Medium voltage/low to medium power) Telephony was the original, and for many years was a driving application for audio amplification. Although a telecoms voice connection is very undemanding compared with modern data communication (it has a very narrow frequency range, typically ~ 300 - 3000 Hz, and also has very poor signal to noise ratio), a specific issue for the telecomms industry was the technique of multiplexing many (up to a thousand) voice lines onto a single cable, at different frequencies.

The advantage of this is a that a single valve "repeater" amplifier can amplify many calls at once, this being very cost effective. The problem is that the amplifiers need to be extremely linear, otherwise "Intermodulation Distortion" (IMD) will result in "crosstalk" between the multiplexed channels. This stimulated development emphasis towards low distortion far beyond the nominal needs of a single voice channel.

Audio

Today the main application for valves is audio amplifiers for high-end Hi-Fi and musical performance use with electric guitars, electric basses, and Hammond organs, although these applications have different requirements regarding distortion which result in different design compromises, although the same basic design techniques are generic and widely applicable to all broadband amplification applications, not only audio.

Post WW-II, the majority of valve power amplifiers are of the Class AB-1 "Push Pull" ultralinear topology, but niche products using the DH-SET and even OTL topologies still exist in small numbers.

Instrumentation Amplifiers

The basic moving coil volt/ammeter itself takes a small current and thus loads the circuit to which it is attached. This can significantly alter the operating conditions in the circuit being measured, clearly an undesirable feature. The Vacuum Tube Volt Meter (VTVM) was developed by taking advantage of the near infinite input impedance of a valve to buffer the circuit being measured from the load of the ammeter. VTVMs have become obsolete since the introduction of the modern Digital Volt Meter (DVM) which typically also has an extremely high input impedance (FET) input.

Valve oscilloscopes share this very high input impedance and thus can be used to measure voltages even in very high impedance circuits. There may typically be 3 or 4 sets of amplification per display channel. In later oscilloscopes, a type of amplificr using a series of tubes connected at equal distances along transmission lines, known as a distributed amplifier was employed to amplify very high frequency vertical signals before application to the display tube. Valve oscilloscopes are now obsolete.

In the closing years of the valve era, valves were even used to make simple "operational amplifiers" - the building blocks of much modern linear electronics. An Op-amp typically has a differential input stage and a totem pole output, the circuit usually having a minimum of five active devices. A number of "packages" were produced that integrated such circuits (typically using two or more glass envelopes) into a single module that could be plugged into a larger circuit (such as an analog computer). Such Valve op-amps were very far from ideal and quickly became obsolete, being replaced with "integrated" (planar silicon) types.

Narrow band/Radio Frequency (RF)/tuned amplifiers

Historically (pre WWII) "transmitting tubes" were among the most powerful tubes available, were usually direct heated by fragile thoriated filaments that glowed like light bulbs. Some tubes were capable of being driven so hard that the anode would itself glow cherry red, the anodes being machined from solid material (rather than fabricated from thin sheet) to be able to withstand this without distorting when heated. Notable tubes of this type are the 845 and 211. Later tetrode/pentodes such as 817 and (direct heated) 813 were also used in large numbers in (especially military) radio transmitters

RF circuits (in particular transmitters, which is something more than simply an amplification gain stage) are significantly different from broadband amplifier circuits. The antenna or following circuit stage typically contains one or more adjustable capacitive or inductive component allowing the resonance of the stage to be accurately matched with carrier frequency in use, to optimize power transfer from and loading on the valve, a so called "tuned circuit".

Broadband circuits often go down to near DC (10 Hz or below) and up to tens or hundreds of kilohertz/low megahertz, and are required to have essentially flat frequency response over this entire range (4 or more orders of magnitude). RF circuits by contrast are typically required to operate over higher frequencies (which makes capacitive and inductive parasitic effects much more of a design challenge) but often a very narrow frequency range. For example, an RF device might be required to operate over the range 144 to 146 MHz (just 1.4% of an octave)

Historically, distortion and out of band emission was less of an issue and Class C could be used. In the 2000s, radio transmitters are overwhelmingly silicon based, even at microwave frequencies (notably consider cellular radio base stations). However an ever decreasing minority of especially high power radio frequency amplifiers (notably for TV) continue to have valve construction. Nevertheless, since the development of radio is inseparable from valve technology, the field of valve amplification remains of historic interest, notably to radio amateurs.

Transistor Amplifiers

The essential role of this active element is to magnify an input signal to yield a significantly larger output signal. The amount of magnification (the "forward gain") is determined by the external circuit design as well as the active device.

Many common active devices in transistor amplifiers are bipolar junction transistors (BJTs) and metal oxide semiconductor field-effect transistors (MOSFETs). Applications are numerous, some common examples are audio amplifiers in a home stereo or PA system, RF high power generation for semiconductor equipment, to RF and Microwave applications such as radio transmitters.

Transistor-based amplifier can be realized using various configurations: for example with a bipolar junction transistor we can realize common base, common collector or common emitter amplifier; using a MOSFET we can realize common gate, common source or common drain amplifier. Each configuration has different characteristic (gain, impedance...).

Operational Amplifiers (op-amps)

An operational amplifier is an amplifier circuit with very high open loop gain and differential inputs which employs external feedback for control of its transfer function or gain. Although the term is today commonly applied to integrated circuits, the original operational amplifier design was implemented with valves.

Fully Differential Amplifiers (FDA)

A fully differential amplifier is a solid state integrated circuit amplifier which employs external feedback for control of its transfer function or gain. It is similar to the operational amplifier but it also has differential output pins.

A fully differential amplifier, usually referred to as an 'FDA' for brevity, is a DC-coupled high-gain electronic voltage amplifier with differential inputs and differential outputs. In its ordinary usage, the output of the FDA is controlled by two feedback paths which, because of the amplifier's high gain, almost completely determines the output voltage for any given input.

The Ideal FDA

For any input voltages the *ideal* FDA has infinite open-loop gain, infinite bandwidth, infinite input impedances resulting in zero input currents, infinite slew rate, zero output impedance and zero noise.

A Real FDA can only approximate this ideal, and the actual parameters are subject to drift over time and with changes in temperature, input conditions, etc. Modern integrated FET or MOSFET FDAs approximate more closely to these ideals than bipolar ICs where large signals must be handled at room temperature over a limited bandwidth; input impedance, in particular, is much higher, although the bipolar FDA usually exhibit superior (i.e., lower) input offset drift and noise characteristics.

Where the limitations of real devices can be ignored, an FDA can be viewed as a Black Box with gain; circuit function and parameters are determined by feedback, usually negative. An FDA as implemented in practice is moderately complex integrated circuit

Limitations of real FDAs

DC imperfections

- Finite gain — the effect is most pronounced when the overall design attempts to achieve gain close to the inherent gain of the FDA.
- Finite input resistance — this puts an upper bound on the resistances in the feedback circuit.
- Nonzero output resistance — important for low resistance loads. Except for very small voltage output, power considerations usually come into play first. (Output impedance is inversely proportional to the idle current in the output stage — very low idle current results in very high output impedance.)
- Input bias current — a small amount of current (typically ~10 nA for bipolar FDAs, or picoamperes for CMOS designs) flows into the inputs. This current is mismatched slightly between the inverting and non-inverting inputs (there is an input offset current). This effect is usually important only for very low power circuits.
- Input offset voltage — the FDA will produce an output even when the input pins are at exactly the same voltage. For circuits which require precise DC operation, this effect must be compensated for.
- Common mode gain — A perfect operational amplifier amplifies only the voltage difference between its two inputs, completely rejecting all voltages that are common to both. However, the

differential input stage of an FDA is never perfect, leading to the amplification of these identical voltages to some degree. The standard measure of this defect is called the common-mode rejection ratio (denoted, CMRR). Minimization of common mode gain is usually important in non-inverting amplifiers (described below) that operate at high amplification.

- Temperature effects — all parameters change with temperature. Temperature drift of the input offset voltage is especially important.

AC Imperfections

- Finite bandwidth — all amplifiers have a finite bandwidth. This is because FDAs use internal frequency compensation to increase the phase margin.
- Input capacitance — most important for high frequency operation because it further reduces the open loop bandwidth of the amplifier.
- Common mode gain
- Noise - all real electronic components (except superconductor) generate noise. You can find devices with 0.8 to several hundreds nv/rtHz noise performance.

Nonlinear Imperfections

- Saturation — output voltage is limited to a peak value, usually slightly less than the power supply voltage. Saturation occurs when the differential input voltage is too high for the op-amp's gain, driving the output level to that peak value.
- Slewing — the amplifier's output voltage reaches its maximum rate of change. Measured as the slew rate, it is usually specified in volts per microsecond. When slewing occurs, further increases in the input signal have no effect on the rate of change of the output. Slewing is usually caused by internal capacitances in the amplifier, especially those used to implement its frequency compensation, particularly using pole splitting.
- Non-linear transfer function — The output voltage may not be accurately proportional to the difference between the input voltages. It is commonly called distortion when the input signal is a waveform. This effect will be very small in a practical circuit if substantial negative feedback is used.

Power Considerations

- Limited output power — if high power output is desired, an op-amp specifically designed for that purpose must be used. Most op-amps are designed for low power operation and are typically only able to drive output resistances down to 2 kΩ.

- Limited output current — the output current must obviously be finite. In practice, most op-amps are designed to limit the output current so as not to exceed a specified level thus protecting the FDA and associated circuitry from damage.

DC Behaviour

Open-loop gain is defined as the amplification from input to output without any feedback applied. For most practical calculations, the open-loop gain is assumed to be infinite; in reality it is obviously not. Typical devices exhibit open-loop DC gain ranging from 100,000 to over 1 million; this is sufficiently large for circuit gain to be determined almost entirely by the amount of negative feedback used. Op-amps have performance limits that the designer must keep in mind and sometimes work around. In particular, instability is possible in a DC amplifier if AC aspects are neglected.

AC Behaviour

The FDA gain calculated at DC does not apply at higher frequencies. To a first approximation, the gain of a typical FDA is inversely proportional to frequency. This means that an FDA is characterized by its gain-bandwidth product. For example, an FDA with a gain bandwidth product of 1 MHz would have a gain of 5 at 200 kHz, and a gain of 1 at 1 MHz. This low-pass characteristic is introduced deliberately, because it tends to stabilize the circuit by introducing a dominant pole. This is known as frequency compensation. Typical low cost, a general purpose FDA exhibits a gain bandwidth product of a few megahertz. Specialty and high speed FDAs can achieve gain bandwidth products of hundreds of megahertz. Some FDAs are even capable of gain bandwidth products greater than a gigahertz.

Video Amplifiers

These deal with video signals and have varying bandwidths depending on whether the video signal is for SDTV, EDTV, HDTV 720p or 1080i/p etc.. The specification of the bandwidth itself depends on what kind of filter is used and which point (-1 dB or -3 dB for example) the bandwidth is measured. Certain requirements for step response and overshoot are necessary in order for acceptable TV images to be presented.

Oscilloscope Vertical Amplifiers

These are used to deal with video signals to drive an oscilloscope display tube and can have bandwidths of about 500 MHz. The specifications on step response, rise time, overshoot and aberrations can make the design of these amplifiers extremely difficult. One of the pioneers in high bandwidth vertical amplifiers was the Tektronix company.

Distributed Amplifiers

These use transmission lines to temporally split the signal and amplify each portion separately in order to achieve higher bandwidth than can be obtained from a single amplifying device. The outputs of each stage are combined in the output transmission line. This type of amplifier was commonly used on oscilloscopes as the final vertical amplifier. The transmission lines were often housed inside the display tube glass envelope.

Distributed amplifiers are circuit designs that incorporate transmission line theory into traditional amplifier design to obtain a larger gain-bandwidth product than is realizable by conventional circuits.

The design of the distributed amplifiers was first formulated by William S. Percival in 1936. In that year Percival proposed a design by which the transconductances of individual vacuum tubes could be added linearly without lumping their element capacitances at the input and output, thus arriving at a circuit that achieved a gain-bandwidth product greater than that of an individual tube. Percival's design did not gain widespread awareness however, until a publication on the subject was authored by Ginzton, Hewlett, Jasberg, and Noe in 1948. It is to this later paper that the term *distributed amplifier* can actually be traced. Traditionally, DA design architectures were realized using valve technology.

Current Rechnology

More recently, III-V semiconductor technologies, such as GaAs and InP have been used. These have superior performance resulting from higher bandgaps (higher electron mobility), higher saturated electron velocity, higher breakdown voltages and higher-resistivity substrates. The latter contributes much to the availability of higher quality-factor (Q-factor or simply Q) integrated passive devices in the III-V semiconductor technologies.

To meet the marketplace demands on cost, size, and power consumption of monolithic microwave integrated circuits (MMICs), research continues in the development of mainstream digital bulk-CMOS processes for such purposes. The continuous scaling of feature sizes in current IC technologies has enabled microwave and mm-wave CMOS circuits to directly benefit from the resulting increased unity-gain frequencies of the scaled technology. This device scaling, along with the advanced process control available in today's technologies, has recently made it possible to reach an fT of 170 GHz and a maximum oscillation frequency (fmax) of 240 GHz in a 90nm CMOS process.

Theory of Operation

The operation of the DA can perhaps be most easily understood when explained in terms of the traveling-wave tube amplifier (TWTA). The DA consists of a pair of transmission lines with characteristic impedances of Z_0 independently connecting the inputs and outputs of several active devices. An RF signal is thus supplied to the section of transmission line connected to the input of the first device. As the input signal propagates down the input line, the individual devices respond to the forward traveling input step by inducing an amplified complementary forward traveling wave on the output line. This assumes the delays of the input and output lines are made equal through selection of propagation constants and lengths of the two lines and as such the output signals from each individual device sum in phase. Terminating resistors Z_g and Z_d are placed to minimize destructive reflections.

The transconductive gain of each device is g_m and the output impedance seen by each transistor is half the characteristic impedance of the transmission line. So that the overall voltage gain of the DA is: where *n* is the number of stages.

Neglecting losses, the gain demonstrates a linear dependence on the number of devices (stages). Unlike the multiplicative nature of a cascade of conventional amplifiers, the DA demonstrates an additive quality. It is this synergistic property of the DA architecture that makes it possible for it to provide gain at frequencies beyond that of the unity-gain frequency of the individual stages. In practice, the number of stages is limited by the diminishing input signal resulting from attenuation on the input line.

Means of determining the optimal number of stages are discussed below. Bandwidth is typically limited by impedance mismatches brought about by frequency dependent device parasitics. The DA architecture introduces delay in order to achieve its broadband gain characteristics. This delay is a desired feature in the design of another distributive system called the distributed oscillator.

Lumped Elements

Delay lines are made of lumped elements of L and C. The parasitic L and the C from the transistors are used for this and usually some L is added to raise the line impedance. Due to the Miller effect in the common source amplifier the input and the output transmission line are coupled. For example for voltage inverting and current amplifying the input and the output form a shielded balanced line.

Due to the current increasing in the output transmission line with every subsequent transistor, less and less L is added to keep the voltage constant and more and more extra C is added to keep the velocity constant. This C can come from parasitics of a second stage. These delay lines do not have a flat dispersion near their cut off, so it is important to use the same L-C periodicity in the input and the output. If inserting transmission lines, input and output will disperse away from each other.

For a distributed amplifier the input is feed in series into the amplifiers and parallel out of them. To avoid losses in the input, no input signal is allowed to leak through. This avoided by using a balanced input and output also known as push-pull amplifier. Then all signals which leak through the parasitic capacitances cancel. The output is combined in a delay line with decreasing impedance. For narrow band operation other methods of phase-matching are possible, which avoid feeding the signal through multiple coils and capacitors. This may be useful for power-amplifiers.

The single amplifiers can be of any class. There maybe some synergy between distributed class E/F amplifiers and some phase-matching methods. Only the fundamental frequency is used in the end, so this is the only frequency, which travels through the delay line version.

Due to this Miller effect a common source transistor acts as a capacitor (non inverting) at high frequencies and has an inverting transconductance at low frequencies. The channel of the transistor has three dimensions. One dimension, the width, is chosen depending on the current needed. The trouble is for a single transistor parasitic capacitance and gain both scale linearly with the width. For the distributed amplifier the capacitance - that is the width - of the single transistor is chosen based on the highest frequency and the width needed for the current is split across all transistors.

Applications

Note that those termination resistors are usually not used in CMOS, but the losses due to these are small in typical applications. In solid state power amplifiers often multiple discrete transistors are used for power reasons anyway. If all transistors are driven in a synchronized fashion a very high gate drive power is needed. For frequencies at which small and efficient coils are available distributed amplifiers are more efficient.

Voltage can be amplified by a common gate transistor, which shows no miller effect and no unit gain frequency cut off. Adding this yields the cascode configuration. The common gate configuration is incompatible with CMOS; it adds a resistor, that means loss, and is more suited for broadband than for high efficiency applications.

- Radio

- Acousto-optic modulator
- Time to digital converter

Microwave Amplifiers

Travelling wave tube (TWT) amplifiers

Used for high power amplification at low microwave frequencies. They typically can amplify across a broad spectrum of frequencies; however, they are usually not as tunable as klystrons.

A traveling-wave tube (TWT) is an electronic device used to amplify radio frequency signals to high power, usually in an electronic assembly known as a traveling-wave tube amplifier (TWTA).

The bandwidth of a broadband TWT can be as high as three octaves, although tuned (narrowband) versions exist, and operating frequencies range from 300 MHz to 50 GHz. The voltage gain of the tube can be of the order of 70 decibels.

The device is an elongated vacuum tube with an electron gun (a heated cathode that emits electrons) at one end. A magnetic containment field around the tube focuses the electrons into a beam, which then passes down the middle of a wire helix that stretches from the RF input to the RF output, the electron beam finally striking a collector at the other end. A directional coupler, which can be either a waveguide or an electromagnetic coil, fed with the low-powered radio signal that is to be amplified, is positioned near the emitter, and induces a current into the helix.

The helix acts as a delay line, in which the RF signal travels at near the same speed along the tube as the electron beam. The electromagnetic field due to the current in the helix interacts with the electron beam, causing bunching of the electrons (an effect called *velocity modulation*), and the electromagnetic field due to the beam current then induces more current back into the helix (i.e. the current builds up and thus is amplified as it passes down). A second directional coupler, positioned near the collector, receives an amplified version of the input signal from the far end of the helix. An attenuator placed on the helix, usually between the input and output helicies, prevents reflected wave from travelling back to the cathode.

Invention, Development and early use

The invention of the TWT is widely attributed to Rudolf Kompfner in 1942–1943, though Nils Lindenblad did patent a device in May 1940 which was remarkably similar to Kompfner's TWT. Kompfner invented the TWT in a British radar lab during World War II. His first sketch of a TWT is dated November 12, 1942, and he built the first TWT in early 1943. The TWT was refined by Kompfner and John Pierce at Bell Labs.

By the sixties TWTs were produced by such companies as the English Electric Valve Company, followed by Ferranti in the seventies.

On July 10, 1962, the first communications satellite, Telstar 1, was launched with a 2 W, 4 GHz RCA-designed TWT transponder used for transmitting RF signals back to the earth. Syncom 2, the first synchronous satellite (Syncom 1 did not reach its final orbit), launched on July 26, 1963 with two 2 W, 1850 MHz Hughes-designed TWT transponders (one active and one spare).

Coupled-Cavity TWT

Helix TWTs are limited in peak RF power by the current handling (and therefore thickness) of the helix wire. As power level increases, the wire can overheat and cause the helix geometry to warp. Wire thickness can be increased to improve matters, but if the wire is too thick it becomes impossible to obtain the required helix pitch for proper operation. Typically helix TWTs achieve less than 2.5 kW output power.

The coupled-cavity TWT overcomes this limit by replacing the helix with a series of coupled cavities arranged axially along the beam. Conceptually, this structure provides a helical waveguide and hence amplification can occur via velocity modulation. Helical waveguides have very nonlinear dispersion and thus are only narrowband (but wider than klystron). A coupled-cavity TWT can achieve 60 kW output power.

Operation is similar to that of a klystron, except that coupled-cavity TWTs are designed with attenuation between the slow-wave structure instead of a drift tube. The slow-wave structure gives the TWT its wide bandwidth. A free electron laser allows higher frequencies.

Traveling-wave tube Amplifier

A TWT integrated with a regulated power supply and protection circuits is referred to as a traveling-wave tube amplifier (abbreviated TWTA and often pronounced “TWEET-uh”). It is used to produce high-power radio frequency signals. The bandwidth of a broadband TWTA can be as high as one octave, although tuned (narrowband) versions exist; operating frequencies range from 300 MHz to 50 GHz.

A TWTA consists of a traveling-wave tube coupled with its protection circuits (as in klystron) and regulated power supply (EPC, electronic power conditioner), which may be supplied and integrated by a different manufacturer. The main difference between most power supplies and those for vacuum tubes is that efficient vacuum tubes have depressed collectors to recycle kinetic energy of the electrons and therefore the secondary winding of the power supply needs up to 6 taps of which the helix voltage needs precise regulation.

The subsequent addition of a linearizer (as for inductive output tube) can, by complementary compensation, improve the gain compression and other characteristics of the TWTA; this combination is called a linearized TWTA (LTWTA, "EL-tweet-uh"). Broadband TWTAs generally use a helix TWT, and achieve less than 2.5 kW output power. TWTAs using a coupled cavity TWT can achieve 15 kW output power, but at the expense of bandwidth.

Uses

TWTAs are commonly used as amplifiers in satellite transponders, where the input signal is very weak and the output needs to be high power. A TWTA whose output drives an antenna is a type of transmitter. TWTA transmitters are used extensively in radar, particularly in airborne fire-control radar systems, and in electronic warfare and self-protection systems. In these types of applications, a control grid is typically introduced between the TWT's electron gun and slow-wave structure to allow pulsed operation. The circuit that drives the control grid is usually referred to as a grid modulator.

Another major use of TWTAs is for the electromagnetic compatibility (EMC) testing industry for immunity testing of electronic devices. A TWT has sometimes been referred to as a "traveling-wave amplifier tube" (TWAT), although this term was never widely adopted. "TWT" has been pronounced by engineers as "twit", and "TWTA" as "tweeta".

Klystrons

Very similar to TWT amplifiers, but more powerful and with a specific frequency "sweet spot". They generally are also much heavier than TWT amplifiers, and are therefore ill-suited for light-weight mobile applications. Klystrons are tunable, offering selective output within their specified frequency range. A klystron is a specialized linear-beam vacuum tube (evacuated electron tube). Klystrons are used as amplifiers at microwave and radio frequencies to produce both low-power reference signals for superheterodyne radar receivers and to produce high-power carrier waves for communications and the driving force for modern particle accelerators.

Klystron amplifiers have the advantage (over the magnetron) of coherently amplifying a reference signal so its output may be precisely controlled in amplitude, frequency and phase. Many klystrons have a waveguide for coupling microwave energy into and out of the device, although it is also quite common for lower power and lower frequency klystrons to use coaxial couplings instead. In some cases a coupling probe is used to couple the microwave energy from a klystron into a separate external waveguide.

All modern klystrons are amplifiers, since reflex klystrons, which were used as oscillators in the past, have been surpassed by alternative technologies. The pseudo-Greek word *klystron* comes from the stem form *klys* of a Greek verb referring to the action of waves breaking against a shore, and the end of the word *electron*.

The brothers Russell and Sigurd Varian of Stanford University are generally considered to be the inventors of the klystron. Their prototype was completed in August 1937. Upon publication in 1939, news of the klystron immediately influenced the work of US and UK researchers working on radar equipment. The Varians went on to found Varian Associates to commercialize the technology (for example to make small linear accelerators to generate photons for external beam radiation therapy). In their 1939 paper, they acknowledged the contribution of A. Arsenjewa-Heil and O. Heil (wife and husband) for their velocity modulation theory in 1935.

During the second World War, the Axis powers relied mostly on (then low-powered) klystron technology for their radar system microwave generation, while the Allies used the far more powerful but frequency-drifting technology of the cavity magnetron for microwave generation. Klystron tube technologies for very high-power applications, such as synchrotrons and radar systems, have since been developed.

Explanation

Klystrons amplify RF signals by converting the kinetic energy in a DC electron beam into radio frequency power. A beam of electrons is produced by a thermionic cathode (a heated pellet of low work function material), and accelerated by high voltage electrodes (typically in the tens of kilovolts). This beam is then passed through an input cavity. RF energy is fed into the input cavity at, or near, its natural frequency to produce a voltage which acts on the electron beam.

The electric field causes the electrons to bunch: electrons that pass through during an opposing electric field are accelerated and later electrons are slowed, causing the previously continuous electron beam to form bunches at the input frequency. To reinforce the bunching, a klystron may contain additional "buncher" cavities. The RF current carried by the beam will produce an RF magnetic field, and this will in turn excite a voltage across the gap of subsequent resonant cavities. In the output cavity, the developed RF energy is coupled out. The spent electron beam, with reduced energy, is captured in a collector.

Two-cavity klystron Amplifier

In the two-chamber klystron, the electron beam is injected into a resonant cavity. The electron beam, accelerated by a positive potential, is constrained to travel through a cylindrical *drift tube* in a straight path by an axial magnetic field. While passing through the first cavity, the electron beam is velocity modulated by the weak RF signal. In the moving frame of the electron beam, the velocity modulation is equivalent to a plasma oscillation.Plasma oscillations are rapid oscillations of the electron density in conducting media such as plasmas or metals.(The frequency only depends weakly on the wavelength). So in a quarter of one period of the plasma frequency, the velocity modulation is converted to density modulation, i.e. bunches of electrons. As the bunched electrons enter the second chamber they induce standing waves at the same frequency as the input signal. The signal induced in the second chamber is much stronger than that in the first.

Two-cavity klystron Oscillator

The two-cavity amplifier klystron is readily turned into an oscillator klystron by providing a feedback loop between the input and output cavities. Two-cavity oscillator klystrons have the advantage of being among the lowest-noise microwave sources available, and for that reason have often been used in the illuminator systems of missile targeting radars. The two-cavity oscillator klystron normally generates more power than the reflex klystron—typically watts of output rather than milliwatts.

Since there is no reflector, only one high-voltage supply is necessary to cause the tube to oscillate, the voltage must be adjusted to a particular value. This is because the electron beam must produce the bunched electrons in the second cavity in order to generate output power. Voltage must be adjusted to vary the velocity of the electron beam (and thus the frequency) to a suitable level due to the fixed physical separation between the two cavities. Often several "modes" of oscillation can be observed in a given klystron.

Reflex Klystron

In the reflex klystron (also known as a 'Sutton' klystron after its inventor), the electron beam passes through a single resonant cavity. The electrons are fired into one end of the tube by an electron gun. After passing through the resonant cavity they are reflected by a negatively charged reflector electrode for another pass through the cavity, where they are then collected. The electron beam is velocity modulated when it first passes through the cavity. The formation of electron bunches takes place in the drift space between the reflector and the cavity.

The voltage on the reflector must be adjusted so that the bunching is at a maximum as the electron beam re-enters the resonant cavity, thus ensuring a maximum of energy is transferred from the electron beam to the RF oscillations in the cavity.The voltage should always be switched on before providing the input to the reflex klystron as the whole function of the reflex klystron would be destroyed if the supply is provided after the input. The reflector voltage may be varied slightly from the optimum value, which results in some loss of output power, but also in a variation in frequency. This effect is used to good advantage for automatic frequency control in receivers, and in frequency modulation for transmitters. The level of modulation applied for transmission is small enough that the power output essentially remains constant. At regions far from the optimum voltage, no oscillations are obtained at all. This tube is called a reflex klystron because it repels the input supply or performs the opposite function of a [Klystron].

There are often several regions of reflector voltage where the reflex klystron will oscillate; these are referred to as modes. The electronic tuning range of the reflex klystron is usually referred to as the variation in frequency between half power points—the points in the oscillating mode where the power output is half the maximum output in the mode. It should be noted that the frequency of oscillation is dependent on the reflector voltage, and varying this provides a crude method of frequency modulating the oscillation frequency, albeit with accompanying amplitude modulation as well. Modern semiconductor technology has effectively replaced the reflex klystron in most applications.

Multicavity Klystron

In all modern klystrons, the number of cavities exceeds two. A larger number of cavities may be used to increase the gain of the klystron, or to increase the bandwidth.

Tuning a Klystron

Some klystrons have cavities that are tunable. Tuning a klystron is delicate work which, if not done properly, can cause damage to equipment or injury to the technician. By adjusting the frequency of individual cavities, the technician can change the operating frequency, gain, output power, or bandwidth of the amplifier. The technician must be careful not to exceed the limits of the graduations, or damage to the klystron can result.

Manufacturers generally send a card with the unique calibrations for a klystron's performance characteristics, that lists the graduations to be set to attain any of a set of listed frequencies. No two klystrons are exactly identical (even when comparing like part/model number klystrons), and so every card is specific to the individual unit. Klystrons have serial numbers on each of them to uniquely identify each unit, and for which manufacturers may (hopefully) have the performance characteristics in a database. If not, loss of the calibration card may be an insoluble problem, making the klystron unusable or perform marginally un-tuned.

Other precautions taken when tuning a klystron include using nonferrous tools. Some klystrons employ permanent magnets. If a technician uses ferrous tools, (which are ferromagnetic), and comes too close to the intense magnetic fields that contain the electron beam, such a tool can be pulled into the unit by the intense magnetic force, smashing fingers, injuring the technician, or damaging the unit. Special lightweight nonmagnetic (aka diamagnetic) tools made of beryllium alloy have been used for tuning U.S. Air Force klystrons.

Precautions are routinely taken when transporting klystron devices in aircraft, as the intense magnetic field can interfere with magnetic navigation equipment. Special overpacks are designed to help limit this field "in the field," and thus allow such devices to be transported safely.

Optical Klystron

In an optical klystron the cavities are replaced with undulators. Very high voltages are needed. The electron gun, the drift tube and the collector are still used.

Floating drift tube Klystron

The floating drift tube klystron has a single cylindrical chamber containing an electrically isolated central tube. Electrically, this is similar to the two cavity oscillator klystron with a lot of feedback between the two cavities. Electrons exiting the source cavity are velocity modulated by the electric field as they travel through the drift tube and emerge at the destination chamber in bunches, delivering power to the oscillation in the cavity. This type of oscillator klystron has an advantage over the two-cavity klystron on which it is based. It only needs one tuning element to effect changes in frequency. The drift tube is electrically insulated from the cavity walls, and DC bias is applied separately. The DC bias on the drift tube may be adjusted to alter the transit time through it, thus allowing some electronic tuning of the oscillating frequency. The amount of tuning in this manner is not large and is normally used for frequency modulation when transmitting.

Collector

After the RF energy has been extracted from the electron beam, the beam is destroyed in a collector. Some klystrons include depressed collectors, which recover energy from the beam before collecting the electrons, increasing efficiency. Multistage depressed collectors enhance the energy recovery by "sorting" the electrons in energy bins.

Applications

Klystrons produce microwave power far in excess of that developed by solid state. In modern systems, they are used from UHF (100's of MHz) up through hundreds of gigahertz (as in the Extended Interaction Klystrons in the CloudSat satellite). Klystrons can be found at work in radar, satellite and wideband high-power communication (very common in television broadcasting and EHF satellite terminals), and high-energy physics (particle accelerators and experimental reactors). At SLAC, for example, klystrons are routinely employed which have outputs in the range of 50 megawatts (pulse) and 50 kilowatts (time-averaged) at frequencies nearing 3 GHz Popular Science's "Best of What's New 2007" included a company using a klystron to convert the hydrocarbons in everyday materials, automotive waste, coal, oil shale, and oil sands into natural gas and diesel fuel.

In St. Petersburg, FL, local cable news station Bay News 9 uses a Klystron Tube in its Klystron 9 Weather Radar to increase its range and resolution. The radar site is actually located behind the Bright House Sports Network building in Pinellas Park, FL.

Confusion with Krytron

A misleadingly similarly named tube, the krytron, is used in simple switching applications. It has recently gained fame as a rapid switch which can be used in nuclear weapons to precisely detonate explosives at high speeds, in order to start the fission process. Krytrons have also been used in photocopiers, raising issues of war technology transfer to countries for items such as this, which have a "dual use."

Musical Instrument (audio) Amplifiers

An audio amplifier is usually used to amplify signals such as music or speech. An instrument amplifier is an electronic amplifier that converts the often barely audible or purely electronic signal from musical instruments such as an electric guitar, an electric bass, or an electric keyboard into an electronic signal capable of driving a loudspeaker that can be heard by the performers and audience. Combination ("combo") amplifiers include a preamplifier, a power amplifier, tone controls, and one or more speakers in a cabinet, a housing usually made of plywood, particleboard, or, less commonly, moulded plastic. Instrument amplifiers for some instruments are also available without an integral speaker; these amplifiers have to be plugged into an external speaker cabinet.

Instrument amplifiers are available for specific instruments, including the electric guitar, electric bass, electric keyboards, and acoustic instruments such as the mandolin and banjo. Some amplifiers are designed for specific styles of music, such as the "traditional"-style "tweed" guitar amplifiers used by blues and country musicians, and the Marshall amplifiers used by hard rock and heavy metal bands.

Unlike home "hi-fi" amplifiers or public address systems, which are designed to reproduce accurately the source sound signals with as little harmonic distortion as possible, instrument amplifiers are often designed to add additional tonal coloration to the original signal or emphasize (or de-emphasize) certain frequencies. The two exceptions are keyboard amplifiers and "acoustic" instrument amplifiers, which typically aim for a relatively flat frequency response.

Types

"Traditional" amplifiers

Standard amps

Standard amplifiers, such as the Fender "tweed"-style amps and Gibson amps, are often used by traditional rock, blues, and country musicians who wish to create a "vintage" 1950s-style sound. They are used by electric guitarists, pedal steel guitar players, and blues harmonica ("harp") players. Combo amplifiers such as the Fender Bassman have tube amplifiers, four 10" speakers, and built-in reverb and "vibrato" effects units.

These amps are designed to produce a variety of sounds ranging from a clean, warm sound (when used in country and soft rock) to a growling, natural overdrive, when the volume is set near its maximum, (when used for blues, rockabilly, and roots rock). These amplifiers usually have a sharp treble roll-off at 5 kHz to reduce the extreme high frequencies, and a bass roll-off at 60–100 Hz to reduce boominess. The nickname "tweed" refers to the lacquered beige-light brown fabric covering used on these amplifiers.

The smallest "combo" amplifiers, which are mainly used for rehearsal and warm-up purposes, may have only a single 8" or 10" speaker. Some harmonica players use these small combo amplifiers for concert performances, though, because it is easier to create natural overdrive with these lower-powered amplifiers. Larger combo amplifiers, with one 12 inch speaker or two or four 10 or 12 inch speakers are used for club performances. For large concert venues, performers may also use an amplifier "head" with several separate speaker cabinets (which usually contain two or four 12" speakers).

Hard Rock and Heavy Metal

These electric guitar amplifiers add an aggressive "drive", intensity, and "edge" to the guitar sound with distortion effects, preamplification boost controls, and tone filters. While many of the most expensive, high-end models use tube amplifiers, there also many models that use transistor amplifiers, or a mixture of the two technologies (i.e., a tube preamplifier with a transistor power amplifier). Amplifiers of this type, such as Marshall amplifiers, are used in a range of the louder, heavier genres of rock, including hard rock, heavy metal, and hardcore punk. This type of amplifier is available in a range of formats, ranging from smaller combo amplifiers for rehearsal and warm-up purposes to heavy "heads" which are designed to be used with separate speaker cabinets, which is colloquially referred to as a "stack."

In the late 1960s and early 1970s, public address systems at rock concerts were used mainly for the vocals. As a result, to get a loud electric guitar sound, early heavy metal and rock-blues bands often used "stacks" of 4x12" Marshall speaker cabinets on the stage. In 1969 Jimi Hendrix used four stacks to create a powerful lead sound, and in the early 1970s by the band Blue Фyster Cult used an entire wall of Marshall Amplifiers to create a roaring wall of sound. In the 1980s, metal bands such as Slayer and Yngwie Malmsteen also used "walls" of over 20 Marshall cabinets. However, by the 1980s and 1990s, most of the sound at live concerts was produced by the sound reinforcement system rather than the onstage guitar amplifiers, so most of these cabinets were not connected to an amplifier. Instead, these walls of speaker cabinets were used for aesthetic reasons.

Amplifiers for harder, heavier genres often use valve amplifiers (known as "tube amplifiers" in North America). Valve amplifiers have a "warmer" tone than those of transistor amps, particularly when overdriven. Instead of abruptly clipping off the signal at cut-off and saturation levels, the signal is rounded off more smoothly. Vacuum tubes also exhibit different harmonic effects than transistors. In contrast to the "tweed"-style amplifiers, which use 10" speakers in an open-backed cabinet, companies such as Marshall tend to use 12" speakers in a closed-back cabinet. These amplifiers usually allow users to switch between "clean" and distorted tones (or a rhythm guitar-style "crunch" tone and a sustained "lead" tone) with a foot-operated switch.

Bass

Bass amplifiers are designed for bass guitars or more rarely, for upright bass. They differ from amplifiers for the regular electric guitar in several respects. They have extended bass response and tone controls optimised for bass instruments, which produce pitches of 40 Hz, in the case of a standard four-string electric bass, or even lower for five- or six-string electric basses.

Higher-end bass amplifiers sometimes include compressor or limiter features, which help to keep the amplifier from distorting at high volume levels, and an XLR DI output for patching the bass signal directly into a mixing board or PA systems. Larger, more powerful bass amplifiers (300 or more watts) are often provided with external metal heat sinks or fans to help keep the amplifier cool.

Speaker cabinets designed for bass instrument amplification usually use larger loudspeakers (or more loudspeakers, in the case of the popular 4 X 10" cabinets, which contain four 10" speakers) than the cabinets used for other instruments, so that they can move the larger amounts of air needed to reproduce low frequencies. While the largest speakers commonly used for regular electric guitar are 12" speakers, electric bass speaker cabinets often use 15" speakers. Bass players who play styles of music that require an extended low-range response, such as death metal, sometimes use speaker cabinets with 18" speakers. The speakers used for bass instrument amplification tend to be more heavy-duty than speakers used for regular electric guitar, and the speaker cabinets are typically more rigidly constructed and heavily braced, to prevent unwanted buzzes and rattles. Bass cabinets often include bass reflex ports or openings in the cabinet, which improve the bass response, especially at high volumes.

Keyboard

This type of amplifier is used to amplify a range of electric and electronic keyboards, such as synthesizers, Hammond organ-style keyboards, stage pianos and electric pianos. Since keyboard instruments contain a wide frequency range, from very low bass notes to extremely high pitches, keyboard amplifiers are often provided with a large woofer speaker to handle the low notes and a horn (or tweeter) for the high notes.

Keyboard amplifiers intended for general use for a range of keyboard applications usually have very low distortion and extended, flat frequency response in both directions. The exception to this rule is keyboard amplifiers designed for the Hammond organ, such as the vintage Leslie speaker cabinet and modern recreations, which have a tube amplifier which is often turned up to add a warm, "growling" overdrive to the organ sound.

Unlike bass amplifiers and electric guitar amplifiers, keyboard amplifers are rarely used in the "amplifier head" and separate speaker cabinets configuration. Instead, most keyboard amplifiers are "combo" amplifiers that integrate the amplifier, tone controls, and speaker into a single wooden cabinet. Another unusual aspect of keyboard amplifiers is that they are often designed with a "wedge" shape, as used with monitor speakers. This allows the cabinet to be rocked back so that it will project sound upwards at a roughly 45' angle, which is more suitable for a seated keyboardist. Keyboard amplifiers often have a simple onboard mixer, so that keyboardists can control the tone and level of several keyboards. In some genres, such as progressive rock, for example, keyboardists may perform with several synthesizers, electric pianos, and electro-mechanical keyboards.

Acoustic

These amplifiers are designed to be used with acoustic instruments such as violin ("fiddle"), mandolin, and acoustic guitar, especially for the way these instruments are used in relatively quiet genres such as folk and bluegrass. They are similar in many ways to keyboard amplifiers, in that they have a relatively flat frequency response, and they are usually designed so that neither the power amplifier nor the speakers will introduce additional coloration.

To produce this relatively "clean" sound, these amplifiers often have very powerful amplifiers (providing from 400 to 800 watts RMS), to provide additional "headroom" and prevent unwanted distortion. Since an 800 watt amplifier built with standard Class AB technology would be very heavy, some acoustic amplifier manufacturers use lightweight Class D amplifiers, which are also called "switching amplifiers."

Acoustic amplifiers are designed to produce a "clean", transparent, "acoustic" sound when used with acoustic instruments with built-in transducer pickups and/or microphones. The amplifiers often come with a simple mixer, so that the signals from a pickup and microphone can be blended. Since the early 2000s, it has become increasingly common for acoustic amplifiers to be provided with a range of digital effects, such as reverb and compression. As well, these amplifiers often contain feedback-suppressing devices, such as notch filters or parametric equalizers.

An audio amplifier is an electronic amplifier that amplifies low-power audio signals (signals composed primarily of frequencies between 20 hertz to 20,000 hertz, the human range of hearing) to a level suitable for driving loudspeakers and is the final stage in a typical audio playback chain. The preceding stages in such a chain are low power audio amplifiers which perform tasks like pre-amplification, equalization, tone control, mixing/ effects, or audio sources like record players, CD players, and cassette players. Most audio amplifiers require these low-level inputs to adhere to line levels. While the input signal to an audio amplifier may measure only a few hundred microwatts, its output may be tens, hundreds, or thousands of watts.

The audio amplifier was invented in 1906 by Lee De Forest when he invented the triode vacuum tube. The triode was a three terminal device with a control grid that can modulate the flow of electrons from the filament to the plate. The triode vacuum amplifier was used to make the first AM radio.

Early audio amplifiers were based on vacuum tubes (also known as *valves*), and some of these achieved notably high quality (e.g., the Williamson amplifier of 1947-9). Most modern audio amplifiers are based on solid state devices (transistors such as BJTs, FETs and MOSFETs), but there are still some who prefer tube-based amplifiers, due to a perceived 'warmer' valve sound. Audio amplifiers based on transistors became practical with the wide availability of inexpensive transistors in the late 1960s.

Design Parameters

Key design parameters for audio amplifiers are frequency response, gain, noise, and distortion. These are interdependent; increasing gain often leads to undesirable increases in noise and distortion. While negative feedback actually reduces the gain, it also reduces distortion. Most audio amplifiers are linear amplifiers operating in class AB.

Filters and Preamplifiers

Historically, the majority of commercial audio preamplifiers made had complex filter circuits for equalization and tone adjustment, due to the far from ideal quality of recordings, playback technology, and speakers of the day.

Using today's high quality (often digital) source material, speakers, and etc., such filter circuits are usually not needed. Audiophiles generally agree that filter circuits are to be avoided wherever possible. Today's audiophile amplifiers do not have tone controls or filters.

Since modern digital devices, including CD and DVD players, radio receivers and tape decks already provide a "flat" signal at line level, the preamp. is not needed other than as volume control. One alternative to a separate preamp. is to simply use passive volume and switching controls, sometimes integrated into a power amp. to form an "integrated" amplifier.

Further Developments in Amplifier Design

For some years following the introduction of solid state amplifiers, their perceived sound did not have the excellent audio quality of the best valve amplifiers. This led audiophiles to believe that valve sound had an intrinsic quality due to the vacuum tube technology itself. In 1972, Matti Otala demonstrated the origin of a previously unobserved form of distortion: transitory intermodulation distortion (TIM), also called slew rate distortion. TIM distortion was found to occur during very rapid increases in amplifier output voltage.

TIM did not appear at steady state sine tone measurements, helping to hide it from design engineers prior to 1972. Problems with TIM distortion stem from reduced open loop frequency response of solid state amplifiers. Further works of Otala and other authors found the solution for TIM distortion, including increasing slew rate, decreasing preamp frequency bandwidth, and the insertion of a lag compensation circuit in the input stage of the amplifier. In high quality modern amplifiers the open loop response is at least 20 kHz, canceling TIM distortion. However, TIM distortion is still present in most low price home quality amplifiers.

The next step in advanced design was the Baxandall Theorem, created by Peter Baxandall in England. This theorem introduced the concept of comparing the ratio between the input distortion and the output distortion of an audio amplifier. This new idea helped audio design engineers to better evaluate the distortion processes within an audio amplifier.

Phonograph (vinyl record) Equalization

Since the mid-1950s, LP phonograph records have been mastered using RIAA equalization, in which the dynamics of the recording have been altered so that the amplitude of the signal that has been cut into the record increases with increasing frequency. Equalization helps to mask the high frequency noise ("hiss") that is generated as the pickup's stylus rubs against the groove walls. The RIAA curve also attenuates the bass, which reduces the maximum excursions of the stylus to a practical level during loud passages. This has the desirable effect of reducing distortion, as well as making the grooves narrower and increasing the potential maximum recording time per record side. Also, with less excursion, less stress is applied to the stylus, which helps to reduce record wear.

During playback, the RIAA curve is reversed by preamplification, resulting in nearly flat frequency response. It should also be noted that the preamplifier is employed to boost the weak signal emitted by a magnetic pickup. Piezoelectric pickups generally produce much higher output voltages and seldom require preamplification.

Prior to the adoption of the RIAA curve, a number of competing and partially incompatible equalization schemes were utilized during record mastering. Early high fidelity systems often had an equalization selector switch to match playback characteristics to the recording curve of the particular label being played. The development and acceptance of the RIAA curve eliminated this requirement.

Applications

Important applications include public address systems, theatrical and concert sound reinforcement, and domestic sound systems. The sound card in a personal computer contains several audio amplifiers (depending on number of channels), as does every stereo or home-theatre system.

Roles

Instrument amplifiers are designed for a different purpose than 'Hi-Fi' (high fidelity) stereo amplifiers used for radios and home stereo systems. Hi-fi home stereo amplifiers are designed to accurately reproduce the source sound signals from pre-recorded music, with as little harmonic distortion as possible.

In contrast, instrument amplifiers are often designed to add additional tonal coloration to the original signal or emphasize certain frequencies. For electric instruments such as electric guitar, the amplifier helps to create the instrument's tone by boosting the input signal gain and distorting the signal, and by emphasizing frequencies deemed to be desirable (e.g., low frequencies) and de-emphasizing frequencies deemed to be undesirable (e.g., very high frequencies).

The two exceptions are keyboard amplifiers and acoustic amplifiers which are used by folk and bluegrass musicians for amplifying acoustic instruments such as acoustic guitar, violin, and mandolin. Acoustic amplifiers typically aim for a relatively flat response across the entire frequency range, much like a Public Address system.

Size and Power Rating

In the 1960s and 1970s, large, heavy, high output power amplifiers were preferred for instrument amplifiers, especially for large concerts, because public address systems were generally only used to amplify the vocals. Moreover, in the 1960s, PA systems typically did not use monitor speaker systems to amplify the music for the onstage musicians. Instead, the musicians were expected to have instrument amplifiers that were powerful enough to provide amplification for the stage and audience. In late 1960s and early 1970s rock concerts, bands often used large stacks of speaker cabinets powered by heavy tube amplifiers such as the Super Valve Technology (SVT) amplifier, which was often used with eight 10" speakers.

However, over subsequent decades, PA systems were substantially improved, and different approaches such as horn-loaded "bass bins" (in the 1980s) and subwoofers (1990s and 2000s) were used to amplify bass frequencies. As well, in the 1980s and 1990s, monitor systems were substantially improved, which allowed sound engineers to provide onstage musicians with a loud, clear, and full-range reproduction of their instruments' sound.

As a result of the improvements to PA systems and monitor systems, musicians in the 2000s no longer need to have huge, powerful amplifier systems; a small combo amplifier patched into the PA will suffice. In the 2000s, virtually all of the sound reaching the audience in large venues comes from the PA system. As well, in the 2000s onstage instrument amplifiers are more likely to be kept at a low volume, because high volume levels onstage makes it harder to control the sound mix and produce a clean sound.

As a result, in many large venues much of the onstage sound reaching the musicians now comes from the monitor speakers, not from the instrument amplifiers. While stacks of huge speaker cabinets and amplifiers are still used in concerts (especially in heavy metal), this is often mainly for aesthetics or to create a more authentic tone. The switch to smaller instrument amplifiers makes it easier for musicians to transport their equipment to performances. As well, it makes concert stage management easier at large clubs and festivals where several bands are performing in sequence, because the bands can be moved on and off the stage more quickly.

Amplifier Technology

Instrument amplifiers may be based on thermionic ("tube" or "valve") or solid state (transistor) technology.

Tube Amplifiers

Vacuum tubes were the dominant active electronic components in amplifiers manufactured from the 1930s through the early 1970s, and tube amplifiers continue to be preferred by some professional musicians. Some musicians believe that tube amplifiers produce a "warmer" or more "natural" sound than solid state units. However, these subjective assessments of the attributes of tube amplifiers' sound qualities are the subject of ongoing debate.

Although tube amplifiers produce more heat than solid state amplifiers, few manufacturers of these units include cooling fans in the chassis. While tube amplifiers do need to attain a proper operating temperature, if the temperature goes above this operating temperature, it may shorten the tubes' lifespan and lead to tonal inconsistencies.

Solid State Amplifiers

By the 1960s and 1970s, semiconductor transistor-based amplifiers began to become more popular because they are less expensive, lighter-weight, and require less maintenance. In some cases, tube and solid state technologies are used together in amplifiers. A common setup is the use of a tube preamplifier with a solid state power amplifier. There are also an increasing range of products that use digital signal processing and digital modeling technology to simulate many different combinations of amp and cabinets.

The output transistors of solid state amplifiers can be passively cooled by using metal fins called heatsinks to radiate away the heat. For high-wattage amplifiers (over 800 watts), a fan is often used to move air across internal heatsinks.

Other amplifier types

Carbon microphone

One of the first devices used to amplify signals was the carbon microphone (effectively a sound-controlled variable resistor). By channeling a large electric current through the compressed carbon granules in the microphone, a small sound signal could produce a much larger electric signal. The carbon microphone was extremely important in early telecommunications; analog telephones in fact work without the use of any other amplifier. Before the invention of electronic amplifiers, mechanically coupled carbon microphones were also used as amplifiers in telephone repeaters for long distance service.

Magnetic Amplifier

A magnetic amplifier is a transformer-like device that makes use of the saturation of magnetic materials to produce amplification. It is a non-electronic electrical amplifier with no moving parts. The bandwidth of magnetic amplifiers extends to the hundreds of kilohertz. The magnetic amplifier (colloquially known as the "mag amp") is an electromagnetic device for amplifying electrical signals.

The magnetic amplifier was invented early in the 20th century, and was used as an alternative to vacuum tube amplifiers where robustness and high current capacity were required. World War II Germany perfected this type of amplifier, and it was used for instance in the V-2 rocket. The magnetic amplifier has now been largely superseded by the transistor-based amplifier, except in a few safety critical, high reliability or extremely demanding applications.

Principle of Operation

Visually a mag amp device may resemble a transformer but the operating principle is quite different from a transformer - essentially the mag amp is a saturable reactor. It makes use of magnetic saturation of the core, a non-linear property of a certain class of transformer cores. For controlled saturation characteristics the magnetic amplifier employs core materials that have been designed to have a specific B-H curve shape that is highly rectangular, in contrast to the slowly tapering B-H curve of softly saturating core materials that are often used in normal transformers.

The typical magnetic amplifier consists of two physically separate but similar transformer magnetic cores, each of which has two windings - a control winding and an AC winding. A small DC current from a low impedance source is fed into the series-connected control windings. The AC windings may be connected either in series or in parallel, the configurations resulting in different types of mag amps. The amount of control current fed into the control winding sets the point in the AC winding waveform at which either core will saturate. In saturation, the AC winding on the saturated core will go from a high impedance state ("off") into a very low impedance state ("on") - that is, the control current controls at which voltage the mag amp switches "on".

A relatively small DC current on the control winding is able to control or switch large AC currents on the AC windings. This results in current amplification.

Applications

Magnetic amplifiers were used extensively as the switching element in early switched-mode (SMPS) power supplies, as well as in lighting control. They have been largely superseded by semiconductor based solid-state switches, though recently there has been some regained interest in using mag amps in compact and reliable switching power supplies. PC ATX power supplies often use mag amps for secondary side voltage regulation.

Magnetic Amplifiers are still used in some arc welders.

Magnetic amplifier transformer cores designed specifically for switch mode power supplies are currently manufactured by several large electromagnetics companies, including Metglas and Mag-Inc. Magnetic amplifiers can be used for measuring high DC-voltages without direct connection to the high voltage and are therefore still used in the HVDC-technique.

Misnomer uses

Late in the 20th century, Robert Carver designed and produced several high quality high powered audio amplifiers, calling them magnetic amplifiers. In fact, they were in most respects conventional audio amplifier designs with an unusual power supply circuit. They were not magnetic amplfiers in the sense of this.

Rotating Electrical Machinery Amplifier

A Ward Leonard control is a rotating machine like an electrical generator that provides amplification of electrical signals by the conversion of mechanical energy to electrical energy. Changes in generator field current result in larger changes in the output current of the generator, providing gain. This class of device was used for smooth control of large motors, primarily for elevators and naval guns. Field modulation of a very high speed AC generator was also used for some early AM radio transmissions.

Johnsen-Rahbek Effect Amplifier

The earliest form of audio power amplifier was Edison's "electromotograph" loud-speaking telephone, which used a wetted rotating chalk cylinder in contact with a stationary contact. The friction between cylinder and contact varied with the current, providing gain. Edison discovered this effect in 1874, but the theory behind the Johnsen-Rahbek effect was not understood until the semiconductor era.

Mechanical Amplifiers

Mechanical amplifiers were used in the pre-electronic era in specialized applications. Early autopilot units designed by Elmer Ambrose Sperry incorporated a mechanical amplifier using belts wrapped around rotating drums; a slight increase in the tension of the belt caused the drum to move the belt. A paired, opposing set of such drives made up a single amplifier. This amplified small gyro errors into signals large enough to move aircraft control surfaces. A similar mechanism was used in the Vannevar Bush differential analyzer.

Optical Amplifiers

Optical amplifiers amplify light through the process of stimulated emission. An optical amplifier is a device that amplifies an optical signal directly, without the need to first convert it to an electrical signal. An optical amplifier may be thought of as a laser without an optical cavity, or one in which feedback from the cavity is suppressed. Stimulated emission in the amplifier's gain medium causes amplification of incoming light. Optical amplifiers are important in optical communication and laser physics.

Laser Amplifiers

Almost any laser active gain medium can be pumped to produce gain for light at the wavelength of a laser made with the same material as its gain medium. Such amplifiers are commonly used to produce high power laser systems. Special types such as regenerative amplifiers and chirped-pulse amplifiers are used to amplify ultrashort pulses.

Doped fibre Amplifiers

Doped fibre amplifiers (DFAs) are optical amplifiers that use a doped optical fibre as a gain medium to amplify an optical signal. They are related to fibre lasers. The signal to be amplified and a pump laser are multiplexed into the doped fibre, and the signal is amplified through interaction with the doping ions. The most common example is the Erbium Doped Fibre Amplifier (EDFA), where the core of a silica fibre is doped with trivalent Erbium ions and can be efficiently pumped with a laser at a wavelength of 980 nm or 1,480 nm, and exhibits gain in the 1,550 nm region.

Amplification is achieved by stimulated emission of photons from dopant ions in the doped fibre. The pump laser excites ions into a higher energy from where they can decay via stimulated emission of a photon at the signal wavelength back to a lower energy level. The excited ions can also decay spontaneously (spontaneous emission) or even through nonradiative processes involving interactions with phonons of the glass matrix. These last two decay mechanisms compete with stimulated emission reducing the efficiency of light amplification.

The *amplification window* of an optical amplifier is the range of optical wavelengths for which the amplifier yields a usable gain. The amplification window is determined by the spectroscopic properties of the dopant ions, the glass structure of the optical fibre, and the wavelength and power of the pump laser.

Although the electronic transitions of an isolated ion are very well defined, broadening of the energy levels occurs when the ions are incorporated into the glass of the optical fibre and thus the amplification window is also broadened. This broadening is both homogeneous (all ions exhibit the same broadened spectrum) and inhomogeneous (different ions in different glass locations exhibit different spectra). Homogeneous broadening arises from the interactions with phonons of the glass, while inhomogeneous broadening is caused by differences in the glass sites where different ions are hosted. Different sites expose ions to different local electric fields, which shifts the energy levels via the Stark effect.

In addition, the Stark effect also removes the degeneracy of energy states having the same total angular momentum (specified by the quantum number J). Thus, for example, the trivalent Erbium ion (Er^{+3}) has a ground state with J = 15/2, and in the presence of an electric field splits into J + 1/2 = 8 sublevels with slightly different energies. The first excited state has J = 13/2 and therefore a Stark manifold with 7 sublevels. Transitions from the J = 13/2 excited state to the J= 15/2 ground state are responsible for the gain at 1.5 μm wavelength. The gain spectrum of the EDFA has several peaks that are smeared by the above broadening mechanisms. The net result is a very broad spectrum (30 nm in silica, typically). The broad gain-bandwidth of fibre amplifiers make them particularly useful in wavelength-division multiplexed communications systems as a single amplifier can be utilized to amplify all signals being carried on a fibre and whose wavelengths fall within the gain window.

Noise

The principal source of noise in DFAs is Amplified Spontaneous Emission (ASE), which has a spectrum approximately the same as the gain spectrum of the amplifier. Noise figure in an ideal DFA is 3 dB, while practical amplifiers can have noise figure as large as 6-8 dB.

As well as decaying via stimulated emission, electrons in the upper energy level can also decay by spontaneous emission, which occurs at random, depending upon the glass structure and inversion level. Photons are emitted spontaneously in all directions, but a proportion of those will be emitted in a direction that falls within the numerical aperture of the fibre and are thus captured and guided by the fibre. Those photons captured may then interact with other dopant ions, and are thus amplified by stimulated emission.

The initial spontaneous emission is therefore amplified in the same manner as the signals, hence the term *Amplified Spontaneous Emission.* ASE is emitted by the amplifier in both the forward and reverse directions, but only the forward ASE is a direct concern to system performance since that noise will co-propagate with the signal to the receiver where it degrades system performance. Counter-propagating ASE can, however, lead to degradation of the amplifier's performance since the ASE can deplete the inversion level and thereby reduce the gain of the amplifier.

Gain Saturation

Gain is achieved in a DFA due to population inversion of the dopant ions. The inversion level of a DFA is set, primarily, by the power of the pump wavelength and the power at the amplified wavelengths. As the signal power increases, or the pump power decreases, the inversion level will reduce and thereby the gain of the amplifier will be reduced. This effect is known as gain saturation - as the signal level increases, the amplifier saturates and cannot produce any more output power, and therefore the gain reduces. Saturation is also commonly known as gain compression.

To achieve optimum noise performance DFAs are operated under a significant amount of gain compression (10 dB typically), since that reduces the rate of spontaneous emission, thereby reducing ASE. Another advantage of operating the DFA in the gain saturation region is that small fluctuations in the input signal power are reduced in the output amplified signal: smaller input signal powers experience larger (less saturated) gain, while larger input powers see less gain.

Inhomogeneous Broadening Effects

Due to the inhomogeneous portion of the linewidth broadening of the dopant ions, the gain spectrum has an inhomogeneous component and gain saturation occurs, to a small extent, in an inhomogeneous manner. This effect is known as *Spectral hole burning* because a high power signal at one wavelength can 'burn' a hole in the gain for wavelengths close to that signal by saturation of the inhomogeneously broadened ions. Spectral holes vary in width depending on the characteristics of the optical fibre in question and the power of the burning signal, but are typically less than 1 nm at the short wavelength end of the C-band, and a few nm at the long wavelength end of the C-band. The depth of the holes is very small, though, making it difficult to observe in practice.

Polarization Effects

Although the DFA is essentially a polarization independent amplifier, a small proportion of the dopant ions interact preferentially with certain polarizations and a small dependence on the polarization of the input signal may occur (typically < 0.5 dB). This is called Polarization Dependent Gain (PDG). The absorption and emission crossections of the ions can be modeled as ellipsoids with the major axes aligned at random in all directions in different glass sites. The random distribution of the orientation of the ellipsoids in a glass produces a macroscopically isotropic medium, but a strong pump laser induces an anisotropic distribution by selectively exciting those ions that are more aligned with the optical field vector of the pump. Also, those excited ions aligned with the signal field produce more stimulated emission.

The change in gain is thus dependent on the alignment of the polarizations of the pump and signal lasers - ie. whether the two lasers are interacting with the same sub-set of dopant ions or not. In an ideal doped fibre without birefringence, the PDG would be inconveniently large. Fortunately, in optical fibers small amounts of birefringence are always present and, furthermore, the fast and slow axes vary randomly along the fibre length. A typical DFA has several tens of meters, long enough to already show this randomness of the birefringence axes. These two combined effects (which in transmission fibers give rise to Polarization Mode Dispersion) produce a misalignment of the relative polarizations of the signal and pump lasers along the fibre, thus tending to average out the PDG. The result is that PDG is very difficult to observe in a single amplifier (but is noticeable in links with several cascaded amplifiers).

Erbium-doped Fibre Amplifiers

The erbium-doped fibre amplifier (EDFA) is the most deployed fibre amplifier as its amplification window coincides with the third transmission window of silica-based optical fibre.

Two bands have developed in the third transmission window - the *Conventional*, or C-band, from approximately 1525 nm - 1565 nm, and the *Long*, or L-band, from approximately 1570 nm to 1610 nm. Both of these bands can be amplified by EDFAs, but it is normal to use two different amplifiers, each optimized for one of the bands.

The principal difference between C- and L-band amplifiers is that a longer length of doped fibre is used in L-band amplifiers. The longer length of fibre allows a lower inversion level to be used, thereby giving at longer wavelengths (due to the band-structure of Erbium in silica) while still providing a useful amount of gain.

EDFAs have two commonly-used pumping bands - 980 nm and 1480 nm. The 980 nm band has a higher absorption cross-section and is generally used where low-noise performance is required. The absorption band is relatively narrow and so wavelength stabilised laser sources are typically needed. The 1480 nm band has a lower, but broader, absorption cross-section and is generally used for higher power amplifiers. A combination of 980 nm and 1480 nm pumping is generally utilised in amplifiers.

The EDFA was invented by a group including David N. Payne, R. Mears, and L. Reekie, from the University of Southampton and a group from AT&T Bell Laboratories, E. Desurvire, P. Becker, and J. Simpson.

Doped Fibre Amplifiers for other Wavelength Ranges

Thulium doped fibre amplifiers have been used in the S-band (1450-1490 nm) and Praseodymium doped amplifiers in the 1300 nm region. However, those regions have not seen any significant commercial use so far and so those amplifiers have not been the subject of as much development as the EDFA. However, Ytterbium doped fibre lasers and amplifiers, operating near 1 micrometre wavelength, have many applications in industrial processing of materials, as these devices can be made with extremely high output power (tens of kilowatts).

Semiconductor Optical Amplifier (SOA)

Semiconductor optical amplifiers are amplifiers which use a semiconductor to provide the gain medium. These amplifiers have a similar structure to Fabry-Perot laser diodes but with anti-reflection design elements at the endfaces. Recent designs include anti-reflective coatings and tilted waveguide and window regions which can reduce endface reflection to less than 0.001%. Since this creates a loss of power from the cavity which is greater than the gain it prevents the amplifier from acting as a laser.

Semiconductor optical amplifiers are typically made from group III-V compound semiconductors such as GaAs/AlGaAs, InP/InGaAs, InP/InGaAsP and InP/InAlGaAs, though any direct band gap semiconductors such as II-VI could conceivably be used. Such amplifiers are often used in telecommunication systems in the form of fibre-pigtailed components, operating at signal wavelengths between 0.85 μm and 1.6 μm and generating gains of up to 30 dB.

The semiconductor optical amplifier is of small size and electrically pumped. It can be potentially less expensive than the EDFA and can be integrated with semiconductor lasers, modulators, etc. However, the performance is still not comparable with the EDFA. The SOA has higher noise, lower gain, moderate polarization dependence and high nonlinearity with fast transient time. This originates from the short nanosecond or less upper state lifetime, so that the gain reacts rapidly to changes of pump or signal power and the changes of gain also cause phase changes which can distort the signals. This nonlinearity presents the most severe problem for optical communication applications. However it provides the possibility for gain in different wavelength regions from the EDFA. "Linear optical amplifiers" using gain-clamping techniques have been developed.

High optical nonlinearity makes semiconductor amplifiers attractive for all optical signal processing like all-optical switching and wavelength conversion. There has been much research on semiconductor optical amplifiers as elements for optical signal processing, wavelength conversion, clock recovery, signal demultiplexing, and pattern recognition.

Vertical-Cavity SOA

A recent addition to the SOA family is the vertical-cavity SOA (VCSOA). These devices are similar in structure to, and share many features with, vertical-cavity surface-emitting lasers (VCSELs). The major difference when comparing VCSOAs and VCSELs is the reduced mirror reflectivities used in the amplifier cavity. With VCSOAs, reduced feedback is necessary to prevent the device from reaching lasing threshold. Due to the extremely short cavity length, and correspondingly thin gain medium, these devices exhibit very low single-pass gain (typically on the order of a few per cent) and also a very large free spectral range (FSR). The small single-pass gain requires relatively high mirror reflectivities to boost the total signal gain. In addition to boosting the total signal gain, the use of the resonant cavity structure results in a very narrow gain bandwidth; coupled with the large FSR of the optical cavity, this effectively limits operation of the VCSOA to single-channel amplification. Thus, VCSOAs can be seen as amplifying filters.

Given their vertical-cavity geometry, VCSOAs are resonant cavity optical amplifiers that operate with the input/output signal entering/exiting normal to the wafer surface. In addition to their small size, the surface normal operation of VCSOAs leads to a number of advantages, including low power consumption, low noise figure, polarization insensitive gain, and the ability to fabricate high fill factor two-dimensional arrays on a single semiconductor chip. These devices are still in the early stages of research, though promising preamplifier results have been demonstrated. Further extensions to VCSOA technology are the demonstration of wavelength tunable devices. These MEMS-tunable vertical-cavity SOAs utilize a microelectromechanical systems (MEMS) based tuning mechanism for wide and continuous tuning of the peak gain wavelength of the amplifier.

Raman Amplifier

In a Raman amplifier, the signal is intensified by Raman amplification. Unlike the EDFA and SOA the amplification effect is achieved by a nonlinear interaction between the signal and a pump laser within an optical fibre. There are two types of Raman amplifier: distributed and lumped. A distributed Raman amplifier is one in which the transmission fibre is utilised as the gain medium by multiplexing a pump wavelength with signal wavelength, while a lumped Raman amplifier utilises a dedicated, shorter length of fibre to provide amplification. In the case of a lumped Raman amplifier highly nonlinear fibre with a small core is utilised to increase the interaction between signal and pump wavelengths and thereby reduce the length of fibre required.

The pump light may be coupled into the transmission fibre in the same direction as the signal (co-directional pumping), in the opposite direction (contra-directional pumping) or both. Contra-directional pumping is more common as the transfer of noise from the pump to the signal is reduced.

The pump power required for Raman amplification is higher than that required by the EDFA, with in excess of 500 mW being required to achieve useful levels of gain in a distributed amplifier. Lumped amplifiers, where the pump light can be safely contained to avoid safety implications of high optical powers, may use over 1W of optical power.

The principal advantage of Raman amplification is its ability to provide distributed amplification within the transmission fibre, thereby increasing the length of spans between amplifier and regeneration sites. The amplification bandwidth of Raman amplifiers is defined by the pump wavelengths utilised and so amplification can be provided over wider, and different, regions than may be possible with other amplifier types which rely on dopants and device design to define the amplification 'window'.

Optical Parametric Amplifier

An optical parametric amplifier allows the amplification of a weak Signal-Impulse in a noncentrosymmetric nonlinear medium (e.g. BBO). In contrast to the previously mentioned amplifiers, which are mostly used in telecommunication environments, this type finds its main application in expanding the frequency tunability of ultrafast solid-state lasers (e.g. Ti:sapphire). By using a noncollinear interaction geometry Optical Parametric Amplifiers are capable of extreme broad amplification bandwidths.

Miscellaneous types

- There are also mechanical amplifiers, such as the automotive servo used in braking.

- Relays can be included under the above definition of amplifiers, although their transfer function is not linear (that is, they are either open or closed).
- Also purely mechanical manifestations of such digital amplifiers can be built (for theoretical, didactical purposes, or for entertainment).
- Another type of amplifier is the fluidic amplifier, based on the fluidic triode.

ELECTRON TUBES

DIODES

The diode is the simplest electron tube. It is often used as a rectifier, that is, a device for changing alternating current into direct current, but recent advances in technology have produced other, nonelectronic rectifiers which are smaller, more efficient, and sometimes cheaper than the electron diode. Diodes will be discussed here prin cipally because they provide an easy approach to the explanation of the more complicated tubes.

The base of the tube is usually made out of some form of plastic material in which prongs are imbedded. A chamber, called the envelope, most often made of glass but sometimes of metal, is cemented into the base. Wires running from the parts inside the envelope are brought out and soldered to the prongs. Therefore, when the tube is plugged into a tube socket, as in the figure, the terminals on the bottom of the socket provide convenient connections to the internal tube parts.

After the elements of the tube are put together, the envelope is evacuated to a high vacuum and sealed off. Inside the envelope are the two basic elements, a source of electrons and a collector. The source can be either of two sorts. The simplest is a filament of tungsten wire coated with various chemicals. When a current is passed through the filament, it heats up and glows, heating the chemical coating on its surface. The chemicals are so constituted that large numbers of electrons are boiled off when they are hot. Thus the coated filament acts as a source of free electrons when current is passed through it.

The other type of electron source is very similar. It consists of a metal surface coated with the same kind of electron-emitting chemicals. The metal is wrapped around an uncoated tungsten filament and is heated when current is passed through the filament. The metal surface is often connected to the filament right inside the tube. The source of electrons is called the cathode because it is usually connected to the negative side of a supply voltage, as will be explained later.

The other important element in the electron diode is the electron collector. It is a relatively large piece of metal (or sometimes carbon) that surrounds the cathode, and is called the plate, or more rarely, the anode.

Suppose that a diode were connected. One source, labeled "*C*," provides the power to heat the cathode, and electrons will boil off into the surrounding space. A second source, labeled *B*, makes the plate positive with respect to the cathode. The electrons that have been freed from the cathode will therefore be attracted toward the plate, and current will flow around the loop indicated by the arrows in the figure. This current is called the plate current. (The current heating the cathode is called the filament or heater current.)

The voltmeter connected across a resistor in the plate circuit will thus register a voltage proportional to the rate at which electrons pass from the cathode to the plate. If a diode is connected, it will exhibit certain properties. First, it may be obvious that the greater the filament current, the greater will be the plate current, because electrons will boil off and be available at a greater rate. There are several limits to this statement, however. Below a critical filament temperature, no electrons at all will be given off. And above some temperature, the filament will melt. Between those limits, the rate of electron emission is a very nonlinear function of filament voltage. For these reasons, the filament voltage in a vacuum tube is almost always held constant, and is not used as a means of varying the plate current.

For a fixed filament voltage, the plate current will vary with the voltage between the filament and the plate. At low plate voltages, electrons are boiled off the cathode more rapidly than they are drawn to the plate, and they tend to accumulate in a cloud around the cathode. This cloud is called the space charge. As the plate voltage increases, more and more of the electrons in the space charge are drawn to the plate. Therefore, the plate current increases and the size of the space charge decreases.

At high plate voltages the electrons are drawn to the plate at the same rate as they are freed from the cathode. Thus the space charge disappears, and further increases in the plate voltage do not increase the plate current appreciably. The preceding discussion applies to the case when the plate voltage is increased from zero to a high positive one, relative to the cathode. When the plate is made negative with respect to the filament, it should be obvious that no current will flow, and the tube is said to be cut off; that is, a diode will conduct current in only one direction, and can therefore be used as a rectifier.

The filament battery simply heats the filament and should be considered as completely separate from the rest of the circuit. Since points *A* and *B* are the two sides of a 115-volt a-c source, *A* will be alternately positive and negative with respect to *B*. When *A* is positive, electrons will flow through the tube and the resistor, but when *A* is negative with respect to *B*, no current will flow. And since, when current does flow, its magnitude will be related monotonically to the magnitude of the voltage, the current through the resistor will change in time. The input voltage is alternating in polarity, and the output voltage and current are pulsating direct current. This circuit is called a half wave rectifier, because only half of the input wave appears in the output.

The circuit is a little more complicated but used much more extensively. It is called a full-wave rectifier circuit, and its action is as follows. The output winding of the transformer has a centre tap which is connected to the cathodes of two separate diodes. (The cathodes may be heated by a second output winding on the transformer, as in this diagram.) When the upper end of the secondary winding *A* is positive with respect to the lower end *B*, it is also true that *A* is positive with respect to the centre tap. Therefore, current will flow through the upper diode, and through the resistor.

At the same time *B* is negative with respect to the centre tap, so the lower diode will not conduct. As soon as the voltage across the secondary reverses, the top diode will stop conducting, but the plate of the lower one will become positive with respect to its cathode, and it will conduct current through the resistor. Thus, whereas on any half-cycle only one of the two diodes is conducting, one or the other will always conduct current through the resistor, and the current will always pass through the resistor in the *same direction* (from *c* to *d*). The voltage and current through the resistor will be as in the diagram. It is said to be full-wave rectified power.

When a current passes through a coil of wire, a magnetic field is generated around the coil, and any change in current causes the field to move, thus inducing a current into any nearby coil. This effect was discussed at length, as it relates to the inducing of a current into the secondary coil of a transformer. However, the changing magnetic field also tends to induce a current into the coil which is generating the field in the first place, and the induced current is always opposite in polarity to the “primary” current, that is, the current that is setting up the field in the first place.

Therefore the net current actually passing through the coil will always be less than it would have been if the magnetic field were not moving. In other words, a coil will conduct more current when a d-c voltage is placed across it than it will with an equivalent a-c voltage. (The property of a coil of wire which gives it greater conductance for direct current than alternating current is called inductance.) Thus, when a voltage is impressed across a coil which has both a-c and d-c components, the coil will tend to block the a-c component more than the d-c component.

The voltage across the resistor is always of the same polarity, but it is pulsating. It can be considered as the sum of a d-c and an a-c component.

If a capacitor is connected across the resistor, it will tend to short-circuit the a-c components while leaving the d-c unaffected for some of the reasons discussed. Thus both capacitors and coils discriminate between a-c and d-c components, and their effects may be combined in what is called a filter circuit, as illustrated at the output of the full-wave rectifier. The output of such a filter has a smoother wave form than the input, and as the filter is made more effective (e.g., by adding more and more components), the output approaches direct current.

The filtering characteristics of various combinations of capacitors, inductances (called chokes), and resistances are too complicated to be discussed here. In general, however, it is important to understand what is meant by the term "filtered" as applied to the output of a power supply. (The development of a loud hum is a trouble that commonly develops in radios and phonographs. The hum is at a frequency of 60 cycles per second, and is caused by the malfunctioning of one or more of the filtering capacitors in the power supply of the amplifier. When a filter capacitor stops working, the voltages to the rest of the radio contain more "60-cycle ripple" than they should, and this results in a 60-cycle modulation of the audio output.)

Full-wave rectifier circuits are so very common as sources of d-c power for electronic apparatus that manufacturers supply special vacuum tubes which contain two diodes within one envelope to save space. Since the two cathodes are always connected together in such a circuit, it is electrically the same thing to use a single cathode and two separate plates. However, for clarity in circuit diagramming, the single cathode in this case is represented as if it were two. The fact that the two diodes are in the same element is indicated by the extension and dotting of the line representing the envelope.

If you examine an actual diode like you may see what looks like another element sticking out to the side of the others. It is also likely that there will be a metallic discolouration of the inside of the envelope near this element. The element is called a "getter." When the tube is being manufactured, the getter is coated with a burnable compound, and after evacuation of the envelope, the coating is ignited to burn out the last bit of oxygen. When the tube is being used, the getter serves no function.

TRIODES

The curve is a simplified plot of the spatial gradient of voltage between the cathode and plate of a conducting diode when the plate voltage is 100 volts. (In an actual diode, the curve dips down below the horizontal axis near the cathode because of the presence of the space charge, but for these purposes, the curve can be considered as always above zero.) If a surface that is transparent to electrons were introduced at the position labeled *G*, and if it were given a voltage of 20 volts negative with respect to the cathode, it would exactly cancel the positive plate voltage between the cathode and *G*. Then electrons leaving the cathode would not be attracted to the plate, and the current through the tube would be cut off.

If the surface were moved closer to the cathode, say to *G'*, then it would require only 10 volts, negative with respect to the cathode, to cut off the current to the plate. Now consider the circuit, where the resistance of *R* is very great compared with the resistance that the diode (without the transparent surface) shows to current flow. When the tube is conducting, almost all of the 100 volts will appear across the resistor, by Ohm's law. If the surface is then introduced at *G'* and is given a voltage of 10 volts, negative with respect to the cathode, the current through the tube and the resistor will be cut off, so the voltage across the resistor will drop to zero. (It is as if the resistance of the tube had been increased to infinity.

The entire plate voltage will appear across the tube.) Thus, the voltage across the resistor has been changed from almost 100 volts to zero by the addition of only 10 volts to the surface at *G'*. Such a setup can then be said to have amplified the input voltage (the voltage between the cathode and the surface) by a factor of ten, considering the voltage across the resistor as the output voltage. The closer the surface to the cathode, the smaller the required cutoff voltage, and the greater the amplification factor.

This sort of amplification is accomplished by inserting a fine wire screen in the position *G'*, directly between the cathode and the plate and much closer to the cathode than to the plate. Such a screen acts essentially as a surface that is transparent to electrons, because most of them travel right through the gaps between wires. But the electric field set up by the screen is fairly uniform, that is, it does not have gaps. It is true, in a real tube, that some of the electrons actually strike the grid, but their number is very small. This kind of tube, containing three basic elements, is called a triode. The screen is called a grid, and when it is designed to operate in the manner just described, it is called a control grid.

For a well-designed triode, as the voltage to the control grid is increased from zero to more negative values with respect to the cathode, the plate current linearly decreases to zero. (The plate voltage supply is held constant in this figure, as it usually is in actual electronic apparatus.) Therefore, if a voltage changing between zero and 10 volts negative is impressed between the grid and the cathode, the changes in voltage across the resistor will have the same shape but will be amplified. The relationship between the input and the output of the amplifier may be represented.

The curve plotted is the same. The output voltage corresponding to any particular input voltage (between cathode and grid) may be found simply by finding the input voltage on the horizontal axis, projecting this up to the curve, and then reading off the corresponding output voltage (plate current multiplied by *R*). Since the output voltage will, in this circuit, change almost instantaneously when the input voltage changes, this procedure may be used to determine the shape of the output wave corresponding to any given input wave. If the input voltage changes abruptly, the output voltage will change proportionally.

That is, so long as the part of the curve being used is linear, the output wave will have the same shape as the input but will be amplified. Here a sine wave is impressed across the input, and a sine wave will appear at the output. You may actually go through the construction of the output sine wave to prove that it is as represented here, but a logical argument may be sufficient to explain the correctness of the plot. It should be clear that any size of input step will be magnified by the same factor at the output. In other words, the shape of the wave containing two unequal steps is faithfully reproduced at the output.

If the sine wave input is considered to be a set of very small steps of differing amplitude, it therefore follows that the output will be a set of larger steps of the same relative sizes as those in the input. Thus the output will also be a sine wave. In fact, any wave form at all will be faithfully reproduced at the output as long as the tube is operating on the linear part of its curve. (This statement holds for the hypothetical amplifier. However, for any real amplifier, a number of properties must be included in the true circuit diagram which limits its capabilities for undistorted reproduction. Some of these properties will be discussed later in this chapter.)

When a so-called grid resistor is added, the circuit will work. The grid resistor is necessary because, if the few electrons that actually do strike the wires of the grid have no path over which to flow back to the cathode, the grid will gradually become more and more negative even with a constant input voltage, and the amplifier will soon cut off completely.

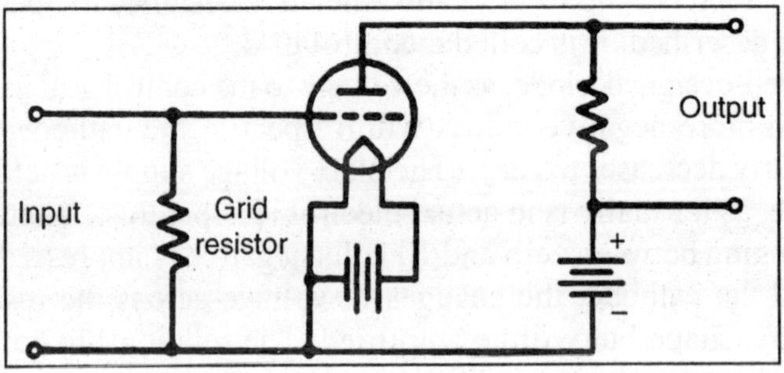

Fig. A Grid Resistor Added to the Triode Amplifier to Prevent the Buildup of Electrons on the Grid.

Should the input to the amplifier drive the grid positive with respect to the plate, the output would become nonlinear. In order to amplify signals of either polarity linearly, a d-c voltage, called the grid-bias voltage, is usually added to the circuit. This voltage sums with the input voltage. Therefore, the grid will remain negative, even when the input voltage is positive, as long as the input voltage is not greater than the grid-bias voltage.

MULTIGRID TUBES

Other grids are added to many vacuum tubes to satisfy special requirements, and the tubes are named accordingly (e.g., a pentode has a filament, a plate, and three grids). The details of these tubes will not be discussed here, except to say that most of them are simply modifications of triodes to perform specialized functions, and they behave according to the same underlying principles.

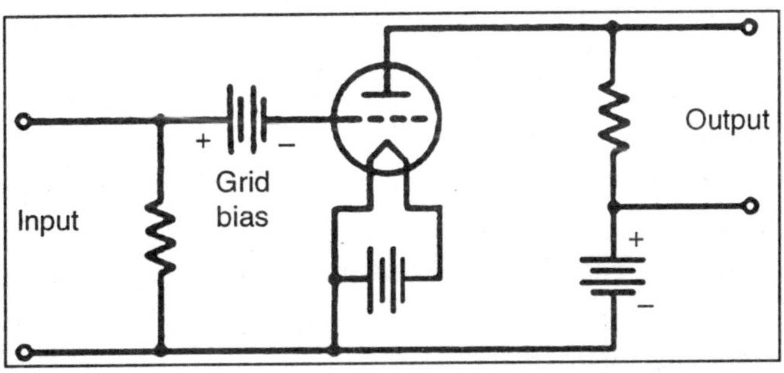

Fig, A Grid-bias Voltage Added to the Triode Amplifier to Insure that the Triode will Operate on the Linear Part of its Curve Regardless of the Polarity of the Input Voltage.

NOMENCLATURE AND SYMBOLS

Vacuum tubes are identified by a set of numbers and letters. Should a designer want to find the exact characteristics of a given tube, he would look it up, according to its numbers and letters, in what is called a tube manual. However, for our purposes, it is sufficient to know only very roughly what the numbers themselves indicate. A very common tube is labeled 6L6, another 117Z5. The 6 and the 117 are the only parts of those labels that really mean anything by themselves. The 6L6 requires a filament voltage of 6 volts and the 117Z5 a filament voltage of 117 volts.

The letters GT after the above kind of designation (e.g., 50L6–GT) refer to the kind of envelope that contains the tube elements. The only difference between a 50L6 and a 50L6–GT is that one is smaller than the other and the GT has a glass envelope instead of the standard metal one. In all but the most specialized equipment, a bad tube of either kind may be replaced by the other as long as there is room for it.

The small, numbered circles refer to the pins on the base of the tube as the tube is viewed from the bottom. When the tube is plugged into a socket, the numbers also refer to the socket looking up from the bottom, and are usually embossed right into the base material next to their respective terminal lugs. Circuit elements soldered to the socket terminal lugs are also visible.

AMPLIFIERS

The action of a simple triode amplifier such has already been described. With a single tube amplifier like this one, an amplification factor of ×80 is about the maximum that can be achieved. However, if a second stage of amplification, just like the first, is simply connected so that its input is the output of the first stage, then the over-all amplification will be the product of the amplification that occurs in each stage considered alone. Thus, the circuit may have an amplification of as much as 6400.

The circuit is in no way affected if the two filaments are connected in parallel across a single power supply, and this is true for most of the circuits that involve electron tubes, as long as the tubes all run at the same filament voltage. However, if the parallel filament connections were all drawn in, they would add needless confusion. Therefore the convention used, terminating the filament leads in *x*'s and showing elsewhere the connection of the *x*'s, is customary in almost all electronic circuit diagrams.

However, for this particular type of amplifier, the plate-voltage supplies cannot be connected in parallel. A separate supply is drawn for each stage in this diagram, and it is a logical necessity of the circuit itself that two supplies are required. The two may be connected together at one end, but there is no way making the circuit work by using the same supply for both stages. If the two plates were connected across a single supply, the amplifier would not work for the following reason.

When there is no current flowing through the first tube, there will be no voltage drop across the resistance R_1. Therefore the grid of the second tube will be very highly positive with respect to its cathode, i.e., the full voltage of the plate supply will be between the grid and cathode. Thus the second tube will conduct very strongly and the amplifier will not work as it should.

CAPACITANCE-COUPLED AMPLIFIERS

The sort of amplifier discussed so far is called a direct-coupled amplifier because the input is connected directly to the first stage, and the first stage is directly connected to the second. For a number of reasons, most of which are too complex to go into here, multistage, direct-coupled amplifiers are very unstable unless a lot of special circuit gimmicks are added. In addition, such amplifiers require a fairly complicated power supply or a lot of batteries, at least one for each stage of amplification, for the reasons discussed in the preceding section. Many of these problems are solved when the input and/or the stages are connected through capacitors. This amplifier is said to be capacitance-coupled. Because a capacitor acts as an extremely high impedance to direct current, virtually all of the steady voltage from the plate supply appears across the capacitor, and the grid of the second tube is not strongly positive with respect to its cathode. Thus a single-plate power supply may be used.

However, for the same reason, only *changes* in the output voltage of the first stage will be impressed upon the second stage grid and appear amplified in the output. In general, the output of a capacitance coupled amplifier is a sort of first derivative of the input because of the action of the capacitor. Thus a square-wave input will appear at the output as a set of spikes, and a sine wave as a sine wave. For a great many purposes, in any audio oscillator or hi-fi amplifier, for example, only changes in the input voltage need be amplified, and capacitance-coupled amplifiers are almost always used in those devices. Most electroencephalographic and nerve-potential amplifiers are also of this type.

For the output, the coupling capacitor has a very small capacitance. Thus it charges up and discharges very rapidly, and the output consists of very short spikes. When the capacitance of the coupling capacitor is increased, it takes longer to charge and discharge. Similarly, if a very-low frequency sine wave is across the input, a small, coupling capacitor will charge up almost as fast as the input to it (it will show a high impedance), and the output voltage will be smaller than if the sine wave had a higher frequency or if the amplifier had a larger coupling capacitance.

In general, any capacitance-coupled amplifier will have a frequency-response curve that shows zero output for a zero frequency input and an increasing output as the frequency increases. Furthermore, the increase in output as a function of the frequency of the input wave will depend upon the value of the coupling capacitance.

The curves drop again at very high frequencies. There are a number of reasons for this which are beyond the scope of this book. However, in general, the following happens. Inside each tube, the grid and cathode and the grid and plate are really capacitors with very low capacitances (they are a pair of conductors separated by an insulating medium). All of the wiring of the circuit itself also has capacitance. As long as the frequency of the input is not too high (e.g., 100,000 cycles per second (cps)), these capacitances have little effect, but when the frequency is really high, they begin to offer relatively low-impedance paths for current to flow, and the circuit acts as if miscellaneous low resistances were put in at undesirable places.

The frequency-response curve of an ideal, *direct-coupled* amplifier would be horizontal from zero to an infinite frequency. For such an amplifier, the output wave could be a perfect, undistorted replica of the input wave for any kind of signal. The frequency response curve for any real amplifier indicates the extent to which the output is distorted.

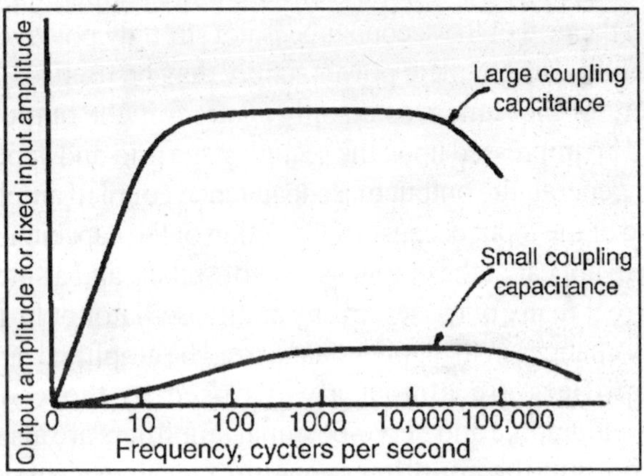

Fig. Frequency Response Curves of a Capacitance-coupled Amplifier for two Different Values of Coupling Capacitance.

If the input is a pure sine wave of any frequency, the output will also be a sine wave of the same frequency, but the amplitude of the output will depend upon the frequency according to the frequency-response curve. (There are causes of distortion not reflected in the frequency-response curve, but they will be ignored in this discussion.)

If the input is any complex wave, it may be considered as the sum of a set of sine waves of various frequencies and amplitudes (Fourier's theorem), and the frequency-response curve indicates how each of these sine components will be attenuated by the amplifier. Thus, a capacitance-coupled amplifier will distort the input wave form by amplifying the low-frequency components of any input wave much less than the high-frequency components.

CHOPPER AMPLIFIERS

Often a slowly changing or steady voltage must be amplified by a large factor, as for example when the psychogalvanic skin response is to be recorded. A multistage, direct-coupled amplifier would serve the purpose, but because such amplifiers must be very carefully designed if they are to be stable, they are relatively expensive. Capacitance-coupled amplifiers are cheaper, but since they cannot amplify low frequencies with the same gain as higher ones, any input signal which contains low-frequency or d-c components will be distorted during amplification.

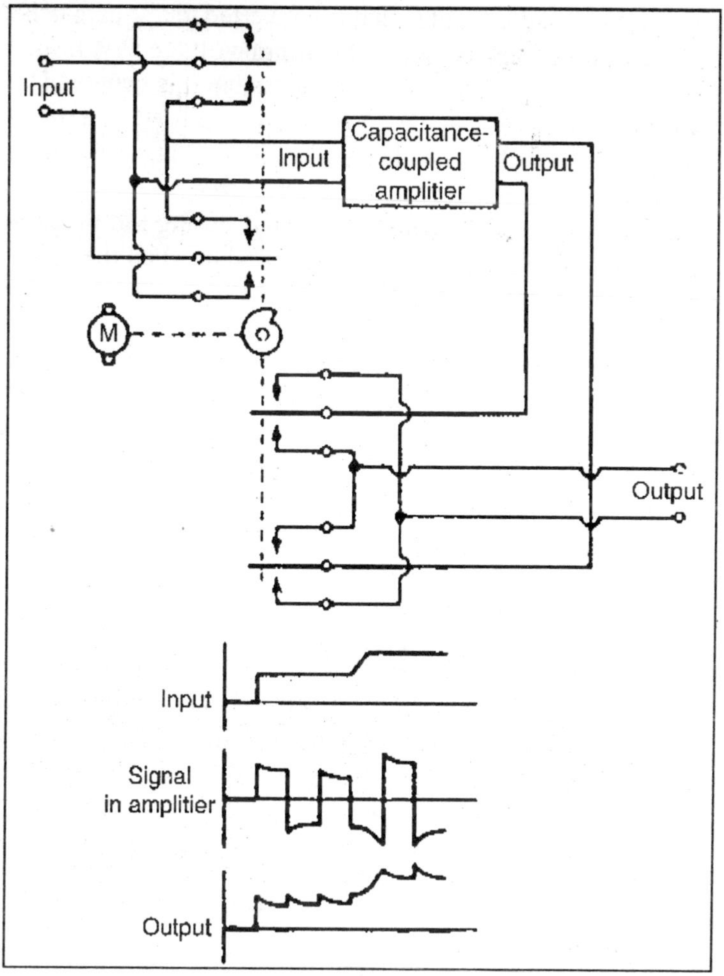

Fig. Semischematic Diagram of a Chopper Amplifier. The Motor Continuously Drives the Cam. Each Rotation of the Cam Reverses the Contact Configuration at the Input and the Output of a Capacitance-Coupled Amplifier. The Input and Output Contact Assemblies are Connected to Act as Reversing Switches.

Amplification of low-frequency signals (including direct current) is therefore sometimes accomplished by a special device called a chopper or breaker amplifier. A chopper amplifier takes a small, steady voltage or a slowly changing one and converts it to a proportional alternating voltage. This a-c signal is fed to a very high-gain, capacitance-coupled amplifier, and then converted back to a proportional d–c signal at the output of the amplifier.

These conversions are sometimes performed by a mechanical chopper. Here the d-c input is fed to a reversing switch which is mechanically driven back and forth. Thus, a d-c input signal is converted to a square wave whose peak-to-trough amplitude is twice the input-voltage level, and whose frequency equals the frequency at which the switch is driven. The square wave is amplified by the capacitance-coupled amplifier, and then fed into another set of contacts that are mechanically connected to the contacts which convert the input.

The output contacts are connected so that during the times the input voltage is normal, the output voltage is normal, but when the input voltage is reversed, the output of the amplifier is reversed, too. Therefore the output of the entire unit is a steady voltage proportional to the input voltage. (It is chopped at the input and "unchopped" at the output.)

The chopping action introduces noise into the output of breaker amplifiers at frequencies near the chopping frequency and higher, so such amplifiers are not suitable for amplifying signals whose frequency sometimes approaches the chopping frequency. It is thus desirable to chop the input at a high rate, and in some chopper amplifiers the mechanical switching has been replaced by a set of electronic switches which chop at rates of 10,000 cps or more.

PUSH-PULL AND GRID-GROUND CONNECTIONS

Most sensitive amplifiers, both direct- and capacitance-coupled, are designed for what is known as push-pull amplification.

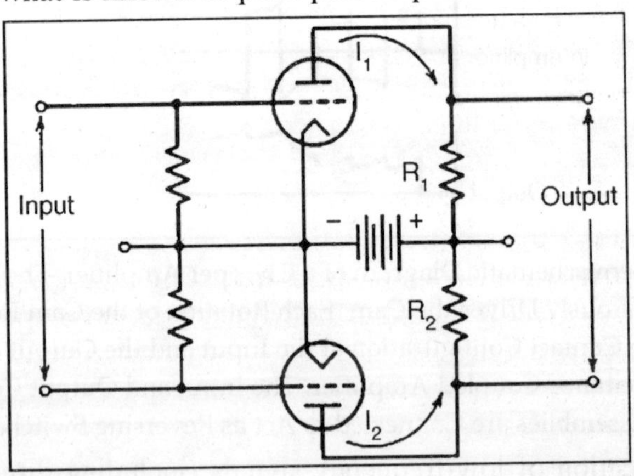

Fig. A One-stage Triode Amplifier with a Push-pull Input and Output.

The input voltage is impressed across a pair of grid resistors and two triodes connected in opposition. When the upper input terminal is positive with respect to the lower one, the upper triode passes more current through the output resistor R_1, and the lower triode less through R_2. Therefore, a difference in voltage appears across the output terminals that are proportional to the input signal. There are several fairly complicated reasons why push-pull circuits are advantageous, and two of them will be mentioned very briefly here. It is often convenient to amplify the algebraic difference between two voltages.

The push-pull circuit automatically does this when each of the two voltages is connected between one of the input terminals and the centre, ground terminal. Suppose, for example, that the occurrence of a stimulus is to be marked directly on a record of a psychogalvanic skin response (PGR). The PGR itself is fed between one of the input terminals and ground, and a pulse at the occurrence of the stimulus is fed between the other terminal and ground. The pulse appears on the record added to the PGR.

Because the push-pull circuit amplifies the algebraic difference between the two input voltages when they are connected, this kind of circuit may be used to reduce unwanted voltages relative to the desired ones. Any two equal voltages that appear with the same polarity at the same time on both sides of the input will cancel each other out, while differences are amplified. One all too common trouble in electronic apparatus is what is called 60cycle pickup. The power lines, lighting circuits, etc., that surround most laboratories actually broadcast fairly strong signals at 60 cps, and any sensitive amplifier will pick these signals out of the air and add them to the voltage that is supposed to be amplified.

For example, if an amplifier is connected to the brain to record brain waves, the record will often show a strong 60-cps wave. When a push-pull input circuit is used, and the central—ground—terminal is also connected to the animal, it is sometimes possible to find a location on the animal where the 60-cycle pickup is equally strong between that location and each of the other two electrodes. When this is the case, the 60-cps pickup is cancelled out, and the uncontaminated brain wave is all that appears at the output of the amplifier.

Most biological amplifiers, and many other amplifiers as well, have push-pull input stages. The input signal may then be connected either from one side to the other, or the signal may be connected between one of the grids and the ground, the other grid being connected directly to the ground. The first of these input connections is called a push-pull input and the second a grid-ground input.

INPUT IMPEDANCE

Consider the measuring of the voltage across a cell membrane. Say that the actual voltage generated is 50 microvolts. The circuit represents a suitable amplifier and a voltmeter. (Actually more than one stage of amplification would be needed, but the addition of other stages to this figure would just complicate things without changing anything basic.) Before the membrane is connected to the amplifier, there is a finite resistance between the two amplifier input terminals, consisting of the grid-to-cathode resistance of the tube in parallel with the grid resistor R_g.

The actual grid-to-cathode resistance of vacuum tubes is so very great that it may be considered infinite for these purposes (and for all but a very few special amplifier circuits). Therefore, the resistance that will be connected from one end of the cell membrane to the other will approximately equal the resistance of R_g. But the cell membrane and the electrodes themselves also have resistance, which may be diagrammed. The current that will flow when the amplifier is connected to the membrane equals $50 \times 10\text{–}6/R_g + R_m$. The voltage that the amplifier will "see" is the voltage between the grid and cathode of its first stage, and this is the voltage across R_g. Therefore, if R_g happens to be equal to R_m, for example, the amplifier will see only 25 microvolts.

In general, as R_g increases relative to the internal resistance of the source and electrodes, the proportion of source voltage that the amplifier will see increases. In addition, as the resistance of the entire input loop increases, the current drawn from the source decreases. If the source happens to be something like a cell, the less current drawn from it, the more normally it will behave. For these reasons, the resistance between the input terminals of an amplifier is important to know, and, at least for biological amplifiers, a large input resistance is desirable.

However, it cannot be made infinitely large because at very high values of Rg (e.g., 40 megohms), the resistances of things such as the insulation between the pins in the base of the tube become comparable to Rg, and the voltage between the grid and cathode is reduced. The term "resistance" has been used in this discussion so far because it is a familiar concept. Actually the more general term "impedance" should be substituted for resistance. The impedance between the input terminals of an amplifier is called the input impedance. For any input signal whose frequency is not too high (say, up to a few thousand cycles per second), the input impedance of a direct-coupled amplifier essentially equals the resistance of the resistor connected between the grid and cathode of the first tube. For higher frequency signals and for capacitance-coupled amplifiers, the calculation or measurement of the input impedance is beyond the scope of this discussion.

OUTPUT IMPEDANCE

Any instrument that is driven by an amplifier (e.g., a meter, relay, etc.) must draw a certain amount of current from the output of the amplifier, and that current must also pass through part of the output stage of the amplifier. The greater the impedance of that part of the amplifier, the more voltage will be lost across it, and the less will be available to drive the meter or relay. The impedance inside the amplifier through which the output current must pass is called the output impedance, and it is exactly analogous to the internal impedance of any power source, such as a battery or a living cell. As a general rule, the lower the output impedance, the better.

Most loudspeakers have fairly low impedances and draw relatively large amounts of current at low voltages, but most vacuum tubes have the opposite property, that is, they pass small currents at high voltages. Therefore vacuum tubes are not very suitable for directly driving loudspeakers. However, the characteristics of tubes and loudspeakers may be matched by placing a transformer between them.

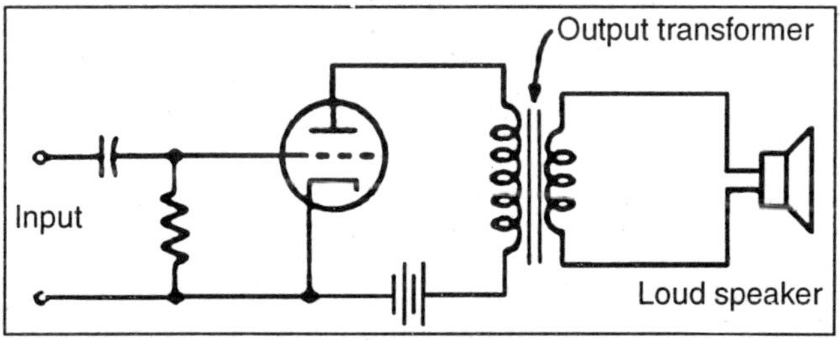

Fig. A One-stage Amplifier Designed to Drive a Loudspeaker. The Output is Transformer-coupled.

The (step-down) transformer is called an output transformer and is used in virtually all vacuum tube amplifiers that are designed to drive loudspeakers. As a rule the secondaries of such transformers have several taps, so that the amplifier may be fitted to speakers with different impedances.

Caution: When the output of an amplifier is transformer-coupled, the amplifier must never be turned on unless the loudspeaker or equivalent resistance is connected across the output terminals. If there is no load (e.g., one of the loudspeaker wires is disconnected), very large voltages may be induced into the output stage of the amplifier, damaging the transformer or output tubes or both.

TRANSISTOR CIRCUITS

When a small signal must be amplified by a very large factor or when the output signal must be an undistorted replica of the input, it is best to buy a commercial amplifier. But there are many occasions when less exacting amplification is required, and for these situations an adequate and cheap transistor amplifier may be easily built.

The fundamental principles of operation of transistors will not be discussed here. They are too complicated and, in many ways, not well enough understood to be worth presenting in this book. What will be discussed is the design and construction of a one-transistor amplifier to be used where a small signal must reliably operate a relay.

Consider, as an example, a situation in which a rat must cross a charged grid to get to food, and the number of times he gets shocked during a 24-hour period is to be counted. Assume that the shock is a constant-current d-c shock of 1 milliampere. If a very sensitive relay, one that closes with a current of less than 1 milliampere, were connected in series with the rat, and the relay were to operate a counter, the counter would register the number of shocks. But relays that sensitive are expensive and hard to find.

If an amplifier were introduced, such as the vacuum-tube amplifier, the signal could be amplified sufficiently to operate a less expensive and more available relay. Although a vacuum-tube amplifier could be built to perform this function, the characteristics of transistors make them particularly suited to this application, and a transistor amplifier would be easier and cheaper to build.

All have three principle elements, an emitter, a base, and a collector. These elements may be considered as analogous to the three elements of a triode vacuum tube. The emitter corresponds to the cathode, the base to the grid, and the collector to the plate. The emitter emits charges which are collected by the collector after having been controlled in density by the base. When the voltage Ec is applied between the emitter and the collector, current will flow through that loop. This current is called the collector current. If the base is then made more positive with respect to the emitter, the collector current will increase, in a way analogous to the increase in plate current when the grid of a triode is made more positive with respect to the cathode. In a vacuum tube, the resistance of the path from the grid to the cathode is extremely high, and virtually no current flows through it. However, the resistance from the base to the emitter of a transistor is relatively low (the actual value depends upon the particular transistor), so the voltage between the base and emitter will cause an appreciable current to flow in that loop.

For a typical transistor, a change of 1 milliampere in the base current will produce a corresponding change of about 40 milliamperes in the collector current. (Common transistors range in amplification factors from about 20 to about 100.) If the transistor is connected into the shock circuit, the 1 milliampere that flows through the rat and the base-emitter circuit will change the current through the relay coil by 40 milliamperes. In this way, the transistor eliminates the need for a very sensitive and expensive relay. The two principle classifications of transistors are the p-n-p and the n-p-n. These letters refer to the physical structures of the transistors themselves. The two types perform in exactly the same way except that all the voltages impressed on a p-n-p must have the opposite polarity from those on an n-p-n. In the above examples, the polarities are correct for the n-p-n transistor. This one was chosen for discussion because the voltages have the same polarities as the corresponding vacuum-tube amplifier voltages. However, many of the transistors most useful for behavioural work are only made with the p-n-p construction.

When the rat is touching the grids, the shock box drives a 1-milliampere direct current through the base-to-emitter circuit of the transistor, making the base negative with respect to the emitter. This base-to-emitter current change produces a corresponding change of about 40 milliamperes in the emitter-collector circuit, which contains the relay coil. The 50,000-ohm variable resistance serves as a bias control. When the slider is all the way to the left, the base is virtually disconnected from the negative side of the battery (because, for this transistor, the base to emitter resistance is considerably lower than 50,000 ohms).

When the slider is all the way to the right, the base is almost the full 22½ volts negative with respect to the emitter. The 1000-ohm fixed resistor prevents the base from going so far negative that the collector current rating is exceeded. The variable resistor may thus be set so that, when there is no shock, the relay has some current flowing through it, but not enough to hold it closed. Then the actual change in collector current need only be big enough to change the relay from open to closed rather than all the way from zero current to closed. The particular transistor is a cheap and versatile one at the time of this writing. However, transistor technology is progressing at such a remarkable rate that it may be considered a collector's item by the time this chapter is printed. The transistor in this circuit is one of a class called power transistors, because they are capable of controlling a large amount of power without overheating. Other types, such as audio-amplifier transistors, are no cheaper and are much more easily damaged by circuit misconnections. For applications in which the output of the amplifier is to control a relay, power transistors are preferable.

Transistors are extremely reliable devices, and are not damaged by fairly severe mechanical shocks (e.g., dropping on the floor). In reliability, size, basic circuit complexity, and efficiency, transistors are superior to vacuum tubes. At their present stage of development, they are inferior to vacuum tubes for two reasons that are revelant to typical behavioural applications. The first important disadvantage of transistors relative to vacuum tubes depends upon the fact that the interelement resistances of transistors are very much lower than the corresponding resistances for vacuum tubes.

Because of these relatively low resistances, the input and output elements and all the stages of a multistage amplifier interact with each other, making the quantitative design of multielement circuits very difficult.

In almost all vacuum-tube amplifiers, for example, the current that flows in the plate circuit of one stage has virtually no effect on the functioning of the preceding stage (unless feedback is deliberately introduced), and each stage can be designed and understood more or less by itself. But changing the value of a resistor anywhere in a multistage transistor circuit changes the functioning of almost everything else in the entire circuit.

The second principle disadvantage of transistors is that they are very sensitive to heat. When soldering a circuit, great care must be taken that the transistors are not made hot. Sometimes a transistor can be firmly connected into a circuit without soldering anything directly to its leads. For example, some transistors fit into sockets. Solderless connections may be made to power transistors. The clips in this photograph are obtained by smashing a tube socket designed to fit a miniature vacuum tube. If wires must be soldered directly to the transistor leads, a pair of pliers or some other heat sink should be clamped between the body of the transistor and the place on the lead where it is to be soldered. Once the circuit is operating, a transistor's sensitivity to heat is manifested in the fact that its parameters (e.g., amplification factor, base-emitter resistance, etc.) change with its temperature.

Therefore any transistor circuit that must be really stable in its characteristics must be carefully designed with special stabilizing circuitry. (Circuits to operate relays need only be designed so that the change in output current is normally well above the minimum necessary to close the relay.) There are many books and booklets on the design of transistor circuits for a variety of applications. Because of the very rapid development of transistor technology, it is not worthwhile to list special references here, but excellent booklets may be purchased at any electronics supply store.

THE CATHODE-RAY OSCILLOSCOPE

The cathode-ray oscilloscope is easily the most versatile piece of electronic equipment to be found in a behaviour laboratory. It may be used directly to measure steady or fluctuating voltages of any level from about 200 microvolts to 1000 volts as well as very short or moderately long time intervals (about 1 microsecond to 50 seconds), to investigate the relationships between two simultaneous signals, to trouble-shoot other electronic apparatus, to provide a visual stimulus that changes in almost any desired way at any time, and even to display moving reversible figures. These are just a few of the functions that may be performed directly and simply with any good, standard laboratory oscilloscope.

This discussion will be based on the assumption that you have an oscilloscope available to experiment with. The oscilloscope should have a direct-coupled vertical amplifier whose gain is calibrated. Also be sure to borrow the oscilloscope instruction manual. Although oscilloscopes can be damaged by mechanical shocks, there is almost no way to damage them electrically (except for one instance which will be explained later). Therefore, anyone who has an oscilloscope available should not hesitate to loan it to you while you learn how to operate it. Do not use an oscilloscope unless it is in its cabinet (there are dangerous voltages within the circuits, but when it is in its cabinet, none of them appear at the front panel). Also avoid any oscilloscope that does not have a plastic or thick glass plate protecting the face of the cathode-ray tube itself. The tube is large and evacuated to a high vacuum. A scratch or blow anywhere on its envelope may cause it to implode.

THE CATHODE-RAY TUBE AND CONTROLS

The cathode-ray tube is the subunit within the oscilloscope that converts electric signals into a corresponding visual display. In the small end is a unit, called the electron gun, consisting of a filament, grids, and plates. Electrons boiled off the filament (or heated cathode) are shaped by the electrostatic field of the grids and plates into a narrow stream moving toward the face of the tube. The inside of the face is coated with a substance, called the phosphor, which gives off light when struck by electrons. The stream of electrons from the electron gun strikes this phosphor layer, and a spot of light is generated which is visible through the glass face of the tube. Usually the phosphor and the surrounding regions on the inside of the tube are coated with a metallic layer which is positive with respect to the cathode, completing the circuit.

Somewhere on the front panel of the oscilloscope, there is a knob labeled "intensity" or "brightness." This knob controls the density of the electron beam and the brightness of the spot. If an intense beam of electrons strikes the phosphor in one small region for a period of time, the phosphor may burn and become discoloured. Therefore, always make sure that, if the spot is stationary on the screen, it is not turned up to a high intensity. Now turn on the power switch of the oscilloscope.

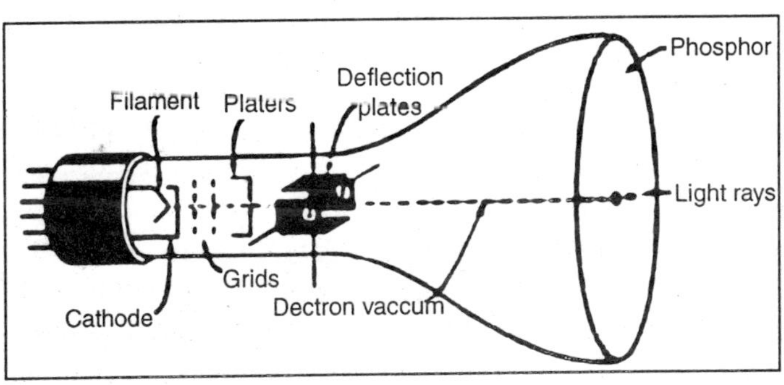

Fig. Semischematic Diagram of a Cathode-ray Tube.

This switch may be a toggle switch, or it may be combined with some other knob on the front panel (e.g., the intensity control). After a minute or two of warmup time, the bright spot should be visible on the face of the cathode-ray tube. If there is no spot, try turning up the intensity control. If this does not work, the spot may be falling somewhere off the screen. There are two control knobs labeled "horizontal (or *x*)," and "vertical (or *y*) position." Turn these knobs until the spot appears. Remember that as soon as it does, the brightness control should be turned down to a moderate level.

If none of these procedures bring anything to the screen, or if a line is present instead of a point, you may have to adjust a different set of controls, called the sweep controls, to be explained later. There is a knob somewhere on the panel which is labeled "horizontal (or *x*) amplifier gain." This control should be set to some low value (e.g., 20 volts per centimeter). If the spot still will not appear, or if it is still a line, look for a control labeled "triggering" or "sweep mode" and set it to a position labeled "horizontal (or *x*) amplifier." If, after all these adjustments are made, you still cannot produce a single stationary spot on the screen, ask the owner to help you.

The intensity control changes the voltages within the electron gun to change the intensity of the electron beam that strikes the screen. The control labeled "focus" changes the relative voltages between elements in the gun to change the diameter of the electron beam as it strikes the screen. Turn the focus control and observe its effect on the spot. Usually the intensity and focus control interact somewhat, so that the spot must be refocused when the intensity is changed.

As the electrons inside the cathode-ray tube stream from the gun to the screen, they pass between two pairs of plates. These plates are called deflection plates. If a voltage is connected between the two plates labeled *x*, in such a way that the left plate is positive with respect to the right one, then, as each electron passes between the plates, it is attracted toward the left plate and repelled from the right one. If the voltage between plates is very great, the electrons may actually strike the left one, but as long as that voltage is low relative to the voltage pulling the electrons toward the screen, the path of each electron will simply be deflected toward the left, but each one will still strike the screen.

Thus the spot on the screen will be deflected to the left. When the horizontal-position control on the front panel of the oscilloscope is turned, it changes the voltage between these two plates in the cathode-ray tube, so that, by turning this control, you can position the spot horizontally on the screen. The plates are called horizontal deflection plates. (Note that the plates themselves are in vertical planes, but they are named according to the direction in which they deflect the spot.)

The cathode-ray tube also contains a second pair of plates which operate in exactly the same way, except that they are in horizontal planes and therefore deflect the spot vertically. The vertical-position control adjusts the voltage between these two vertical deflection plates.

SWEEP GENERATOR AND CONTROLS

The unit labeled "sweep generator" will be discussed now. The sweep generator is an electronic device which generates a sawtooth wave form. The output voltage of this device is as plotted, and the pattern is repeated each time the sweep generator is triggered. The output of the sweep generator may be connected (through an amplifier) to the horizontal deflection plates by means of a switch on the front panel of the oscilloscope. Now suppose that the spot is positioned at the left side of the screen and the connected sweep generator is triggered.

The voltage between the horizontal deflection plates will slowly and linearly increase from zero to some maximum value and then drop to zero again. The spot will therefore move across the screen smoothly until voltage drops to zero, at which point it will jump back to its initial position. In a properly designed cathode-ray tube, the deflection of the spot is a linear function of the voltage between the deflection plates. Therefore, as long as the output of the sweep generator is linear, the spot will move linearly across the screen.

On the front panel of the oscilloscope there is a cluster of controls related to the sweep generator. One of these controls may be switched to a position labeled "free-running." (The control itself is probably labeled "sweep mode.") When this control is set to freerunning, the sweep generator is automatically triggered every time its output voltage drops to zero. In other words, its output will be a series of saw-teeth.

Set the control to the free-running position, and observe that the spot is now in continuous motion. (You may also have to change another switch in the same cluster to a position labeled "sweep.") If the sweep happens to be set to a slow enough speed, you will be able to see the spot slowly move across the screen from left to right, then jump back and start again. If the sweep is set to a very high speed, the spot will move so fast that it appears as a horizontal line. Find the control labeled "sweep speed." It should be calibrated in terms of seconds, milliseconds, and microseconds per centimeter. This control changes the slope of the saw-tooth wave. When the control is turned to increase the slope, two things happen to the motion of the spot.

First, it moves more rapidly across the screen because the voltage between the horizontal deflection plates is increasing more rapidly. Second, so long as the sweep circuit is free-running, the repetition rate will increase for the following reason. The sweep generator circuit is so designed that the voltage will drop back to zero when it reaches some predetermined level. When the slope is greater, that level will be reached sooner, and each sweep will occupy a shorter time. Further, when the circuit is free-running, it automatically retriggers itself each time the voltage drops to zero.

Therefore the repetition rate will increase as the slope of the saw-tooth increases. Change the setting of the sweep speed control and observe its effects upon the display. Many of the controls associated with the sweep circuit are involved in the manner in which the sweep is triggered (when it is not in the free-running state), and these controls will be discussed later.

THE VERTICAL AMPLIFIER

The vertical and horizontal deflection plates of a well-designed cathode-ray tube have independent effects upon the position of the spot. If a voltage is applied to the vertical deflection plates at the same time as another is applied to the horizontal ones, the spot will move up or down as well as horizontally. On the front panel of the oscilloscope is a group of two or three terminals labeled "vertical (or *y*) input." These terminals are usually placed at the lower left side of the panel, and are arranged. They are the input terminals of an amplifier, called the vertical amplifier, whose output is impressed directly across the vertical deflection plates of the cathode-ray tube. One of them is grounded (connected to the cabinet of the oscilloscope and the cathodes of the amplifier). If there is only one other terminal in this group, it is internally connected to the grid of the first tube in the vertical amplifier. If there are two nongrounded terminals in the group, they lead one to each of the two grids of a push-pull input stage. One of these two terminals may be connected externally to the grounded terminal, permitting the amplifier to be used with a grid-ground input.

Now suppose a sine wave is impressed between the grounded and one of the nongrounded input terminals (the third terminal being externally connected to ground). If there is no voltage across the horizontal deflection plates, the spot will move up and down with a sinosoidal motion, and if the frequency of the sine wave is great enough, the display will appear as a vertical line. Switch the sweep circuit back to the condition in which a stationary spot is displayed.

Then connect a wire at least 5 feet long to the nongrounded terminal and turn up the gain of the vertical amplifier until a vertical line appears on the screen. (Leave the other end of the wire disconnected.) All the a-c power lines in the building actually broadcast 60-cps radiation throughout the nearby space. The wire will act as an antenna, picking up some of this radiation and introducing it as a 60-cps voltage between the terminal and ground. This voltage, amplified by the vertical amplifier, thus drives the spot up and down 60 times a second.

Now if the spot is simultaneously moved horizontally across the screen, the shape of the 60-cps wave will be visible. Display it in this way by twisting the horizontal position control rapidly back and forth. If the spot is moved horizontally with a displacement directly proportional to time (i.e., with a constant velocity), the display will be a plot of the 60-cps voltage on the vertical axis against time on the horizontal axis. Switch the sweep circuit back to free-running again, so that the spot is given a constant horizontal velocity, and the vertical wave shape should be visible.

It is unlikely that the display you now see is very informative. First, the sweep speed will probably not be optimal. Adjust it until just one or two complete cycles are visible. Now the display will probably be a ragged sine wave moving horizontally across the screen. The voltage actually picked up by the antenna has this ragged wave form. The raggedness is produced by miscellaneous equipment drawing power from the 60-cps source. The horizontal drift of the display occurs because the sweep is not synchronized with the 60-cps vertical signal. This point requires some clarification. If the period of the horizontal saw-tooth were precisely some even multiple of the period of the wave on the vertical axis, the sweep would always begin and end at the same point on the vertical wave, and the wave would be stationary on the screen.

When there is a slight difference between the periods of the saw-tooth and the vertical signal, each sweep begins at a different point with respect to the vertical wave, and the wave seems to move across the screen. By making a fine adjustment of the sweep speed control, it may be possible to make the wave stop drifting for short periods of time, but the typical sweep generator is not stable enough to hold its period constant over long times. Therefore the wave will begin to move again.

When repetitive waves are to be examined and measured, it is important that they be stationary on the screen, and special circuits are built into all oscilloscopes to eliminate display drift. Some oscilloscopes have what is called a "synch" control. When this control is adjusted correctly, some of the vertical input signal is fed to the input of the sweep circuit, causing it to trigger in synchronism with the vertical signal. If your oscilloscope has such a control, you may prevent the drifting of the display by first setting the sweep speed control to minimize drift, and then turning the synch control until the wave seems to lock in place.

TRIGGERING

Each single saw-tooth wave and sweep delivered by the sweep circuit is triggered when the voltage across the input to the sweep circuit reaches some particular level. This voltage is called the trigger voltage. A selector switch, or a set of them, on the control panel of the oscilloscope allows the operator to choose any one of four different trigger voltage sources. These choices are labeled "free running," "external," "line," and "internal."

External Triggering

In the "external" position, the input to the sweep circuit is connected internally between the ground and another terminal on the control panel. This terminal is usually labeled "ext" or "ext trigger." With the switch in this position, a single sweep will be generated each time the voltage between the external input terminal and the ground reaches some particular level. Suppose, for example, that you wish to observe the nerve impulses resulting from electrical stimulation of a nerve fibre. Electrodes on the fibre are led through an external amplifier to the vertical input terminals of the oscilloscope, and part of the *stimulating* voltage is connected between the external trigger and a ground terminal.

Now, each time the shock is delivered, the sweep is simultaneously triggered. When the sweep speed is properly adjusted, and if the latency of the impulses is constant, the impulses will always appear at the same place on the screen regardless of when the stimulus happens to be delivered. And if, for example, the sweep speed is 1 millisecond per centimeter and the gap between the beginning of the sweep and the first impulse is 2 centimeters, the latency is 2 milliseconds.

This external trigger control may be used anytime you wish to observe the vertical input during some specified period of time, as long as the start of the period can be signaled by a voltage large enough to trigger the sweep. The actual voltage required for triggering is usually a few volts, but the level varies among oscilloscopes.

Internal Triggering

When the trigger selector is turned to "internal," the signal delivered to the vertical deflection plates is also connected across the trigger circuit. Therefore, whenever the vertical input signal reaches some particular voltage, a single sweep is triggered. Switch the trigger selector on your oscilloscope to "internal." The display should now appear as a stationary picture of the vertical input signal since the sweep begins each time at the same point on the vertical wave.

If there is no sweep, and the display is simply a vertical line, find the control labeled "trigger level" and turn it until triggering occurs. (Also be sure that the gain of the vertical amplifier is great enough to give a vertical deflection of several centimeters.)

The trigger-level control adjusts the voltage level of the vertical signal that just triggers the sweep. As you turn this control, the displayed wave will appear to move across the screen. What is actually happening is that the place on the wave where the sweep begins is being changed by the trigger-level control.

Associated with the trigger-level control is a switch usually labeled "trigger slope" or "polarity." The vertical signal passes through the triggering level twice each cycle, once while it is increasing and again as it is decreasing. The position of the "slope" control determines whether the sweep will begin when the vertical signal is increasing (+slope) or decreasing (–slope).

The occurrence of the sweep, when it is internally triggered, depends not upon the input voltage to the vertical amplifier but rather on the output voltage of that amplifier (the input voltage to the vertical deflection plates). Therefore, as the gain of the vertical amplifier is changed, the occurrence of the sweep will shift with respect to the wave itself, beginning earlier as the gain increases. The sweep really begins when the vertical wave, *as it appears on the screen*, reaches some level on the screen, and that level is determined by the setting of the trigger level control.

The internal triggering mode is the one most often used in general laboratory work. If the signal to be examined is repetitive and regular, this mode will hold the display fixed on the screen, and the signal can easily be magnified, vertically with the vertical gain control, and horizontally (along the time dimension) with the sweep speed control.

For example, the wave shape, frequency, and amplitude of a flashing light may be measured simply by connecting a phototube directly to the vertical input terminals when the sweep is in the internally triggered mode. Events which are not repetitive may also be observed in this sweep mode by adjusting the trigger level control to be very sensitive. The sweep will then begin very shortly after the event begins (e.g., when the vertical voltage rises one millimeter above the zero level on the screen). To illustrate the use of the oscilloscope in measuring the duration of a single event, set your sweep circuit to "internal triggering," the sweep speed to about 5 centimeters per second, and the vertical amplifier to approximately 5 volts per centimeter.

Now connect one side of your 22½-volt battery to one of the oscilloscope terminals and briefly touch the other battery lead to the other terminal. When the trigger level control is properly set, the sweep will begin when the lead is first connected, and the trace will then proceed across the screen, dropping vertically when the lead is disconnected. The distance traveled before the trace drops divided by the sweep speed equals the duration of voltage applied to the oscilloscope.

LINE TRIGGERING

With the trigger mode control in the "line" position some of the 60-

cps line voltage is fed to the trigger circuit. Each time the line voltage reaches some level, the sweep is triggered. The level at which triggering occurs may be adjusted with the trigger level control. This position is used whenever a signal which happens to be synchronized with the line voltage is to be observed. The principal use of line triggering in behavioural research is to observe 60-cps pickup.

This pickup often appears along with the signal to be measured, and acts as noise. When the trigger selector is set on "line," this noise component may be observed easily because it is fixed on the screen while the remaining signal components are not (unless they happen to be synchronized with the line voltage). Measures may then be taken to reduce the 60 cps component.

FREE RUNNING

When the sweep generator is free-running, the cessation of each sawtooth automatically triggers the next one, so that the sweep is continuous. This mode is useful during the initial adjustment of the vertical gain and other controls because the trace is continuously visible.

In the other sweep modes, the sweep only occurs once each time the trigger level is reached. In the internal mode, for example, once the vertical signal has triggered a single sweep, another sweep cannot occur until the vertical signal drops below the triggering level and then reaches it again.

If the vertical signal were to stay above the triggering level, no more sweeps would occur. Therefore the free-running mode is also used when the level of a d-c signal is to be measured. Measure the voltages of your battery with the oscilloscope. First set the sweep to free-running and the height of the line to some convenient level (with the vertical position control). Next set the vertical gain to a low value (e.g., 10 volts per centimeter). Now connect the voltage to be measured between the vertical input terminals.

The line on the screen should move up or down, depending upon the polarity of the voltage. If it moves when the connection is first made, but then drifts back to its initial level, make sure that the vertical amplifier is direct-coupled. (There is a switch somewhere near the other vertical amplifier controls that is labeled "a–c" and "d–c."

In the d-c position, the input terminals are directly coupled to the input stage of the vertical amplifier. In the a–c position, an internal capacitor is connected between the nongrounded input terminal and the grid of the first amplifier stage.)

There are two knobs with which to control the vertical amplifier gain—a coarse and a fine control. The coarse control should be calibrated (volts per centimeter), and there is a position of the fine-control knob for which the coarse-control calibrations are correct. If you are not sure where that position is, refer to the oscilloscope manual.

THE HORIZONTAL AMPLIFIER

The horizontal sweep circuit discussed in the foregoing is used whenever the characteristics of a signal are to be examined as a function of time; the horizontal component of spot motion is linear with time. However, there are many applications in which it is desirable to display the vertical signal as a function of some other signal not linearly related to time. To produce this kind of display, the sweep circuit is disconnected from the horizontal deflection plates of the cathode-ray tube.

Then the signal that is to form the horizontal component is fed to the input of the "horizontal" (or *x*) amplifier, and the output of that amplifier is connected to drive the horizontal deflection plates. To demonstrate this procedure, follow these steps:

- Reconnect the length of wire to the nongrounded vertical input terminal.
- Find the horizontal input terminals. They usually have the same configuration as the vertical input terminals, and are usually at the lower right side of the control panel. Connect another length of wire to the nongrounded horizontal input terminal. Do not connect the two lengths of wire together.
- Find the switch that disconnects the sweep circuit and connects the horizontal amplifier output across the horizontal deflection plates. This switch may be the same one that selects trigger modes, or it is sometimes an extreme position of the coarse-gain control on the horizontal amplifier. When this switch is in the correct position, the display will be a diagonal line or oval. If it is a vertical line, increase the gain of the horizontal amplifier, and if it is almost horizontal, decrease the gain.

If the vertical and horizontal deflection plates were receiving identical signals at all instants in time, the display would necessarily be a line inclined at 45 degrees to the horizontal, since, for any given vertical displacement (*y* value), the horizontal displacement (*x* value) would be identical. In another language, if the correlation between the vertical and horizontal signals is perfect, all the points will fall on a regression line at 45 degrees. When the two signals are different in amplitude only, the angle of the line will change accordingly, and if they are also different in phase, the display will become elliptical.

If you have an audio oscillator available, try the following entertaining and instructive demonstration. If you cannot borrow an oscillator, follow the first step in the demonstration anyway.

- Impress the 60–cps line voltage across the vertical amplifier input. There are several ways to do this, but it will be good practice for you to follow the particular procedure to be described. One of the terminals of the wall socket is internally connected to the ground (i.e., water pipes, cement floor, etc.) and the other is 115 volts "above" ground. Furthermore, one of the oscilloscope input terminals is "grounded," which means that it is connected to the oscilloscope cabinet. If you were to connect the nongrounded wall terminal to the grounded oscilloscope terminal, 115 volts would appear between the cabinet and the floor, water pipes, etc., subjecting anyone who touches the cabinet to a severe shock hazard.

But as long as the grounded wall terminal is connected to the grounded oscilloscope terminal and the "high" side of the wall socket to the nongrounded input terminal, the shock hazard is not present (unless you touch the nongrounded input terminal). To find the side of the wall socket that is grounded, connect the grounded side of the oscilloscope to a water pipe or other ground (the metal plate covering the wall socket will also serve as a ground). Then connect the nongrounded input terminal to first one and then the other wall socket terminal.

When a 115–volt sine wave appears on the screen, the oscilloscope is connected to the nongrounded side of the wall, and when no vertical signal is displayed, it is connected to the grounded side. (The same test may be performed with an ordinary light bulb. Connect one terminal of the bulb to a water pipe or the wall socket cover, and the other terminal of the bulb to one of the wall terminals. The bulb will light when it is connected to the nongrounded side of the wall socket.) When the grounded wall terminal is connected to the grounded input terminal, and the nongrounded wall and input terminals are also connected together, a good (nonragged) sine wave should appear on the screen.

- Connect the output terminals of the oscillator to the horizontal input terminals of the oscilloscope. If one of the oscillator terminals is marked with a ground symbol, it should be connected to the grounded oscilloscope input terminal, but if neither oscillator terminal is so marked, either may be connected to the grounded oscilloscope terminal. (There are a number of grounded terminals on the oscilloscope, and, since all are connected together, they all are equivalent. For example, the oscillator may be connected between the nongrounded horizontal input terminal and the grounded vertical input terminal.)
- Set the oscilloscope so that the horizontal amplifier is driving the horizontal deflection plates (sweep-disconnected), set the

oscillator to approximately 60 cps, and turn up the gains of the oscillator and horizontal amplifier until a two-dimensional pattern appears on the screen. Then adjust the frequency of the oscillator until the pattern is a simple line or oval and is as stationary as possible. When the pattern is stationary, the oscillator will be generating a sine wave of exactly 60 CPS. Now set the oscillator to approximately 120 CPS and observe the pattern.

When this double loop is stationary, the oscillator output is exactly 120 CPS. This procedure may be followed for various multiples of 60 CPS, and the oscillator calibrated in this way. The figures generated are called Lissajou figures. Notice that, as they move, they appear to rotate in the third dimension, and the direction of apparent rotation changes from time to time. This kind of display is a moving "reversible figure," analogous to the wellknown Necker cube, staircase, etc., reversible figures.

Independent control of the vertical and horizontal components of spot movement allows the oscilloscope to be used for the presentation of all sorts of moving visual stimuli. For example, suppose human reaction time to a silent visual stimulus is to be measured. The circuit will permit this measurement to be taken. Initially, the subject sees a stationary spot on the screen. When the experimenter closes his key, a clock is started and the spot instantaneously moves horizontally to a new position. The subject is instructed to press his key as soon as the spot moves, and his key stops the clock. The circuit allows the spot to be moved from its initial position to any other point on the screen, the new location being determined by the settings of the two potentiometers and reversing switches. (The reversing switches change the direction of movement.)

The spot may be made to move around the screen according to any predetermined pattern by using the simple principle illustrated in the following example. Let us arbitrarily decide to make the spot move up and down according to the pattern (say, as a stimulus for complex tracking behaviour). If a voltage varying in this way as a function of time were impressed across the vertical input, the spot would move as desired. To generate such a voltage, all that is necessary is a photocell, a light bulb, and a device to drive a paper belt. The desired pattern is drawn as a line on a strip of opaque paper, and the paper is cut along the line. The cut pattern is then formed into a belt and driven continuously through the device.

In this device, rays from the bulb pass by the belt and through a slit. Some of them strike the belt, whereas others go past it and fall on a photocell. As the pattern on the belt moves past the slit, the amount of light reaching the photocell changes as a linear function of the belt width. Photocells will be discussed in more detail, but to understand this apparatus, all that one needs to know is that a vacuum phototube, when connected in the circuit, has an output that is linear with the incident light intensity. Therefore, as the belt moves, the spot will move on the screen with the desired motion.

Z-AXIS MODULATION

Somewhere on the control panel or on the back of the oscilloscope there is a terminal labeled "Z axis." This terminal is connected to one of the elements in the cathode-ray tube in such a way that a voltage impressed between it and any of the grounded terminals will change the brightness of the spot. Unfortunately, in almost all oscilloscopes, the connection is a capacitance coupling. Only *changes* in the applied voltage will affect the spot brightness. The faster the applied voltage is increasing, the brighter will be the spot, and the faster it is decreasing, the dimmer will be the spot.

The spot therefore cannot be extinguished or brightened for any but very short times. Z-axis modulation is most commonly used to introduce a time marker into the trace, particularly when the display is being photographed. A voltage pulse applied periodically between the Z-axis terminal and ground will generate a periodic dotting of the trace.

Phosphors

There are several different kinds of phosphors used to coat the insides of the faces of cathode-ray tubes, and the characteristics of these materials vary along two principal dimensions. Different phosphors generate light spots of different colours and persistences. The phosphors are designated by the letter *P* and a number. The following is a list and evaluation of the three most common ones. A complete list may be found in any booklet on cathode ray tubes, available at most electronic suppliers.

P1—Green Spot, Medium Persistence

This phosphor generates most of its light in the spectral region to which human eyes are most sensitive. The spot is green, and it continues to glow for a short time after the electron beam has moved to a new region of the screen. It is often used for displays that are to be visually observed (instead of photographically recorded). However, in behavioural research, the visual efficiency of the spot on the screen is rarely of any consequence, and since the *P*1 phosphor has other shortcomings, it is not to be recommended.

P7—Short Persistence Blue, Long Persistence Yellow

In the author's experience, this phosphor has proved much more useful than any of the others. When the electron beam is actually striking a point on the screen, the spot is blue.

As soon as the beam leaves the point, the blue disappears (short persistence, time constant approximately 0.5 millisecond) and is replaced by a yellow afterglow which persists for a long time. The blue radiation is well matched to the spectral sensitivity of most films, and thus may be easily photographed, while the yellow is well matched to human visual sensitivity, and lasts long enough that one may actually measure the characteristics of a single sweep with a ruler after the sweep has been completed.

When a trace is being photographed with moving film, a longpersistence glow can sometimes produce a sort of smearing of the record, and, for this reason short persistence is desirable. The shortpersistence blue of the *P*7 phosphor is so much more actinic than the yellow for most films that the yellow afterglow will not produce smearing. However, if the film happens to be sensitive to yellow, or if the yellow afterglow is undesirable for some other reason, a colour filter that passes the blue but cuts out the yellow light may be placed over the screen. Oscilloscope manufacturers sometimes supply such a filter with the cathode ray tube.

P11—Blue, Short Persistence

This phosphor has desirable characteristics when the oscilloscope is to be used to obtain photographic records of very fast phenomena (e.g., the shapes of nerve impulses). For this application, it is superior to the *P*7 for two reasons.

First, the spot is brighter and can therefore be photographed when it is moving at a greater speed. Second, the persistence is somewhat shorter than that of the blue P7 spot, thus smearing is reduced. However, because of its short persistence, events that do not occur at high and regular repetition rates are very hard to observe visually on a *P*11 screen.

When ordering a new oscilloscope, you may specify the particular phosphor you want, since any cathode-ray tube may be coated with any of the common phosphors. Should you already have an oscilloscope with an unsuitable phosphor, you may replace the cathode-ray tube with one containing a more appropriate phosphor. A cathoderay tube to fit any given oscilloscope, and with any common phosphor, may be ordered from the manufacturer of the oscilloscope. Instructions for changing cathode-ray tubes are included in the oscilloscope manual.

POWER SUPPLIES

Most laboratory equipment is powered by the standard 115–volt, 60cps line voltage. Should some a–c voltage other than the line voltage be required, a transformer can be used between the wall socket and the instrument. However, there are many devices that run on direct current, and these require a special power supply. Batteries may be used if very little current is drawn from them, but they are often inconvenient. Power supplies are now being manufactured which convert the wall power into direct current of almost any power and stability that is needed.

Power supplies may be grouped into a number of categories. Highvoltage power supplies (from about 100 to several thousand volts) are usually designed to deliver relatively small amounts of current, in the milliampere range, and are used to supply power to electronic circuits. Low-voltage power supplies (up to 100 volts) are available to deliver currents as great as 50 amperes. In each of these groups, the units may be regulated or not. If a power supply simply rectifies the alternating current from the wall and filters it, it is not regulated. If, however, it contains circuits which maintain the same output voltage regardless of changes in the wall voltage or changes in the amount of current drawn from the unit, it is said to be regulated.

Power supplies, therefore, are rated on each of the following properties:

- Input voltage (usually 115 volts, 60 cps).
- Output voltage (may be fixed, adjustable from zero to some maximum, or adjustable from a minimum not zero to a maximum).
- Output current (the maximum amount that may be drawn without damage. The actual current drawn at any instant will depend on the voltage and the resistance of the unit getting the power).
- Amount of ripple (that is, the amplitude of the a-c component which gets past the filtering and into the output. This is usually stated as a percentage of the output voltage).
- Line regulation (the change in output voltage resulting from some particular change in the input voltage).
- Load regulation (the change in output voltage resulting from a change in the output current).

Before the advent of transistors, power supplies which delivered low voltages and high currents could not be very well regulated. For this reason, whenever it was required that the power to an element be extremely stable (e.g., the current through a bulb in a visual threshold experiment), batteries were used. However, modern transistorized power supplies are even more stable than batteries for applications in which currents greater than I milliampere or so are required.

ELECTRONIC TIMERS

Almost all of the cheaper electronic devices to deliver controlled time intervals, and some of the more expensive ones, operate on exactly the same principles as those of the timers discussed.

The temporal characteristics of the charging or discharging of a resistance-capacitance circuit are used to control a relay or set of them. Commercial timers usually contain a power supply, a set of capacitors and resistors with calibrated time constants, and an amplifier. The amplifier allows the use of more rugged and reliable relays than the sensitive relays necessary for the circuits.

Time intervals are selected by switches on the front panel, which connect combinations of capacitors and resistors to give different time constants. This sort of timer is available to deliver intervals of from about 0.01 second to about 100 seconds.

The reliability of the intervals depends very much more upon the particular make of timer. The chief source of interval variability in these timers is poor regulation of the power supply. But since most manufacturers fail to report the regulation of their built-in power supply, it is hard to select a timer on that basis.

Should a timer prove unreliable, it is sometimes possible to improve it by running it from a constant-voltage transformer. A more expensive, but far more reliable, class of timers may be built up from three basic elements—a constant frequency oscillator, a counter, and a device that operates a relay and/or sends out a pulse when the count reaches some predetermined value. Each of these elements may be bought separately, but there are commercial units that contain all three.

This sort of timer starts counting the cycles of the oscillator when a switch is closed or a voltage is placed across its input, and closes a relay or delivers a voltage pulse when a number that has been preset on its control panel is reached.

Therefore, its accuracy is as good as the constancy of its oscillator (plus or minus onehalf the period of oscillation). Since it is fairly easy to construct a highly stable oscillator, the accuracy of the timer is usually 0.1% or better, plus or minus one-half period. If the internal frequency is 10,000 cps, for example, the time interval will never be off by more than 0.1% of the preset interval ±0.05 millisecond. The same unit may be used to *measure* time intervals with equally good accuracy. The unit is started with one event, stopped with another, and the count, readable on the front panel, indicates the time between events.

Counting timers may be purchased with any number of preset units working off the same counter. For example, a timer with six presets may be used to control the time of occurrence of six different events. These units may also be connected to recycle automatically when any preset count is reached, so a sequence may be repeated over and over.

Chapter 10

Electrical Transients

The discussion will centre around a specific laboratory problem-the registration of the activity of an animal. This problem is a common one, and is representative of problems which involve the sensing of transients. It has a great number of solutions, and several pertinent ones will be explained. To state the problem more specifically, suppose we wish to measure the diurnal variations in the activity of an animal confined to a living cage.

When the animal moves around, he exerts forces on the cage. Probably the easiest way to sense the animal's movement would be to suspend the cage on springs and detect the movement of the cage. In the old days, your assistant would have set about to smoke a kymograph drum and to rig up a system of levers, strings, and sealing wax to scratch a crooked line in the smoke. Next he would have varnished the record to preserve it, and after it had dried he would have spent many painful hours measuring the characteristics of the crookedness of the line. In selecting what to measure, he would have followed very specific instructions based on decisions that you had made previously. Now these instructions are crucial, and their choice and phrasing involve some very subtle considerations.

For example, in order to get a rough measure of gross activity, he might have been told to count all the times the record reverses directions, or to count the number of times that the trace abruptly changes its angle. This sort of instruction can be qualified more and more to meet specific objectives. The reason this is being discussed in a book on electric apparatus is that any piece of apparatus designed to replace your assistant must embody the same instructions that would have been given to him. The nature of these instructions and of the sensing system itself determines the structure of the apparatus. Conversely, practical considerations about the structure of the apparatus often influence the nature of the instructions.

INDUCTIVE SENSING

The apparatus to be designed is supposed to measure activity, and activity consists of *changes* in the animal's posture or position, independent of the posture or position itself. If the cage were suspended by a spring, changes around the resting level would indicate the animal's movements. First a device will be designed to follow the instruction, "Count the number of times the cage moves faster than some threshold rate." This instruction suggests the use of some electric device that responds only to transients.

Here a permanent magnet fastened rigidly to the cage moves in and out of a coil of wire when the cage is moved. A voltage proportional to the velocity of the cage will be generated between the terminals of the coil. At zero velocity, there will be zero voltage, etc. This magnet-coil combination is a simple and very effective sensor of movement. If a very sensitive relay were connected across the coil and an electromagnetic counter were operated from the relay contacts, a count would be registered each time the animal moved at a velocity greater than some threshold velocity. The threshold speed will be determined by the strength of the magnet, the number of turns on the coil, and the sensitivity of the relay. The voltage divider in the circuit serves as an over-all sensitivity control.

In practice, would require an extremely sensitive relay for reasonable operation, so that a simple amplifier should probably be interposed between the sensing coil and the relay coil. Its inclusion in the circuit does not change the basic principles of operation of this activity sensor. The circuit might not always follow the intended instructions. For example, it would probably give only one count for a motion, even though the movement changed direction several times.

Although the direction of current reverses on each cycle and the current magnitude thus goes through zero for an instant at each reversal, the inertia of the relay and counter would probably be sufficient to prevent the relay from ever opening. Electromagnetic counters are built so that they register one count for each make *and* break of current through their coils. If current is continuous, the counter does not continuously advance (counters function very much like stepping relays).

The circuit avoids this difficulty. The magnet coil is connected to the relay through a rectifier. A rectifier is a device which passes current easily in one direction but shows a high resistance to current flow in the opposite direction. Therefore, the relay operates only on motion in one direction. Just for closure, the circuit is included. This circuit will automatically record activity over a 24-hour period, storing the sum of all activity in each successive 2-hour period on a separate counter.

TRANSFORMERS AS TRANSIENT SENSORS

Each movement of the cage changes the current flowing through the relay coil, and the values can be chosen so that, every time the cage moves downward farther than some threshold position, the relay closes. For example, suppose that the supply voltage is 10 volts, and that the relay is one which closes when the voltage across its coil reaches 5 volts. Then the relay will close and the counter will add a count whenever the cage moves the slider downward past the mid-point of the resistor. In general, the total count will equal the total number of times the slider passed this threshold position while moving downward.

There is a serious difficulty with this kind of activity recording system. The count produced by a fixed amount of activity will depend upon the animal's weight. Consider an animal whose weight happens to be such that, when he is at rest, the slider is just at the threshold level. If he now engages in activity which bounces the cage up and down irregularly, the counter will register every bounce. If that animal is now replaced by one that is lighter, so that the resting position of the slider is above the threshold level, when he moves, some of the bounces will not be big enough to move the slider beyond the threshold position, and they will not register.

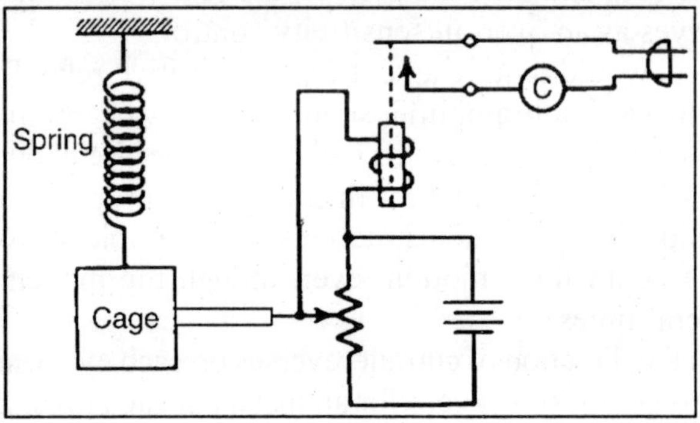

Fig. Circuit to Register a Count Each Time the Cage Moves Downward Past Some Threshold Position.

Thus the same amount of actual activity will result in a smaller count. For the same reason, any change in a single animal's weight, such as would result from eating or urinating, would be indistinguishable from a change in his activity. This defect can be remedied by putting, between the voltage divider and the relay, a device that senses only electric transients. The transformer functions in the same way in this circuit as when connected in the usual way. Any current flowing through the primary coil sets up a magnetic field in the region of the secondary coil.

As long as the primary current is steady, there is no movement of the field with respect to the secondary coil, and no voltage is induced into it. But every change in the input current induces a voltage in the secondary. The greater the rate of change of current in the primary, the greater the amplitude of the output voltage. For these reasons, the circuit will register a count each time the animal moves faster than some threshold rate. In this case too, it is usually more practical to insert a simple amplifier between the transformer and the relay. Otherwise an extremely sensitive relay is needed.

Another common situation in which a circuit must discriminate between transients and steady states is the one in which the making and/or breaking of a circuit is to be sensed. For example, in Skinner boxes, usually one and only one reinforcement is given for each bar press. If the animal sits on the bar, the reinforcement should not continue to pour out. In other words, the circuit that controls the reinforcement should operate only when the response circuit is changing from off to on. There are numerous ways to accomplish this, and the one best suited to any given problem depends on other related apparatus elements.

Here, each time the bar is pressed, the amount of current flowing through the primary of the transformer increases from zero to some finite value. A magnetic field is thus built up and, while it is building, it generates enough voltage in the secondary to close the relay. As soon as the current in the primary reaches equilibrium, the voltage in the secondary again drops to zero and the relay opens.

Therefore, even if the bar is held down for a long time, the relay stays closed for only a short time. When the bar is released, the magnetic field in the transformer collapses, generating another pulse of voltage in the secondary, and this pulse might also be expected to deliver a reinforcement. However, the current generated in the secondary of the transformer flows in opposite directions for the making and the breaking of the primary circuit. Therefore the rectifier, when connected, will prevent the relay from closing when the bar is released.

POWER QUALITY

In its broadest sense, power quality is a set of boundaries that allows electrical systems to function in their intended manner without significant loss of performance or life. The term is used to describe electric power that drives an electrical load and the load's ability to function properly with that electric power. Without the proper power, an electrical device (or load) may malfunction, fail prematurely or not operate at all. There are many ways in which electric power can be of poor quality and many more causes of such poor quality power.

The electric power industry comprises electricity generation (AC power), electric power transmission and ultimately electricity distribution to an electricity meter located at the premises of the end user of the electric power. The electricity then moves through the wiring system of the end user until it reaches the load. The complexity of the system to move electric energy from the point of production to the point of consumption combined with variations in weather, generation, demand and other factors provide many opportunities for the quality of supply to be compromised.

While "power quality" is a convenient term for many, it is the quality of the voltage - rather than power or electric current - that is actually described by the term. Power is simply the flow of energy and the current demanded by a load is largely uncontrollable.

It is often useful to think of power quality as a compatibility problem: is the equipment connected to the grid compatible with the events on the grid, and is the power delivered by the grid, including the events, compatible with the equipment that is connected? Compatibility problems always have at least two solutions: in this case, either clean up the power, or make the equipment tougher.

Ideally, voltage is supplied by a utility as sinusoidal having an amplitude and frequency given by national standards (in the case of mains) or system specifications (in the case of a power feed not directly attached to the mains) with an impedance of zero ohms at all frequencies.

No real-life power source is ideal and generally can deviate in at least the following ways:

- Variations in the peak or RMS voltage are both important to different types of equipment.
- When the RMS voltage exceeds the nominal voltage by 10 to 80% for 0.5 cycle to 1 minute, the event is called a "swell".
- A "dip" (in British English) or a "sag" (in American English - the two terms are equivalent) is the opposite situation: the RMS voltage is below the nominal voltage by 10 to 90% for 0.5 cycle to 1 minute.
- Random or repetitive variations in the RMS voltage between 90 and 110% of nominal can produce a phenomenon known as "flicker" in lighting equipment. Flicker (i.e. light flicker) is the impression of unsteadiness of visual sensation induced by a light stimulus on the human eye. A precise definition of the voltage fluctuations (i.e. voltage flicker) that produce light flicker that annoys humans, has been subject to ongoing debate in more than one scientific community for many years.
- Abrupt, very brief increases in voltage, called "spikes", "impulses", or "surges", generally caused by large inductive loads being turned off, or more severely by lightning.
- "Undervoltage" occurs when the nominal voltage drops below 90% for more than 1 minute. The term "brownout" is an apt description for voltage drops somewhere between full power (bright lights) and a blackout (no power - no light). It comes from the noticeable to significant dimming of regular incandescent

lights, during system faults or overloading etc., when insufficient power is available to achieve full brightness in (usually) domestic lighting. This term is in common usage has no formal definition but is commonly used to describe a reduction in system voltage by the utility or system operator to decrease demand or to increase system operating margins.

- "Overvoltage" occurs when the nominal voltage rises above 110% for more than 1 minute.
- Variations in the frequency
- Variations in the wave shape - usually described as harmonics
- Nonzero low-frequency impedance (when a load draws more power, the voltage drops)
- Nonzero high-frequency impedance (when a load demands a large amount of current, then stops demanding it suddenly, there will be a dip or spike in the voltage due to the inductances in the power supply line)

Each of these power quality problems has a different cause. Some problems are a result of the shared infrastructure. For example, a fault on the network may cause a dip that will affect some customers and the higher the level of the fault, the greater the number affected, or a problem on one customer's site may cause a transient that affects all other customers on the same subsystem.

Other problems, such as harmonics, arise within the customer's own installation and may or may not propagate onto the network and so affect other customers. Harmonic problems can be dealt with by a combination of good design practice and well proven reduction equipment.

Power Conditioning

Power conditioning is modifying the power to improve its quality. An uninterruptible power supply can be used to switch off of mains power if there is a transient (temporary) condition on the line. However, cheaper UPS units create poor-quality power themselves, akin to imposing a higher-frequency and lower-amplitude sawtooth wave atop the sine wave.

A surge protector or simple capacitor or varistor can protect against most overvoltage conditions, while a lightning arrestor protects against severe spikes. Electronic filters can remove harmonics.

Smart Grids and Power quality

Modern systems use sensors called PMUs distributed throughout their network to monitor power quality and in some cases respond automatically to them.

Using such smart grids features of rapid sensing and automated self healing of anomalies in the network promises to bring higher quality power and less downtime while simultaneously supporting power from intermittent power sources and distributed generation, which would if unchecked degrade power quality.

CAPACITATIVE SENSING

Any electric current can be considered as made up of two components, a steady or d-c one and a changing or a-c one. Such components are clear in the voltage pattern, where a battery is connected in series with the wall-socket voltage. An animal sitting in an activity cage and twitching might generate a similar wave form.

In the preceding discussion, the transformer has been described as a transient sensing device. But it is also possible, and in some ways more useful, to consider the transformer as a device that transmits the a-c components of a signal while blocking the d-c component.

Exactly the same thing can be done with an altogether different circuit element—the capacitor (condenser, *archaic*). A capacitor consists of two sheets of a conductive material separated by a good insulator. Assume that the capacitor is electrically neutral, that is, there are neither too few nor too many electrons on the plates or the glass. Now suppose that the two terminals of the capacitor are connected to the terminals of a battery. The negative terminal of the battery has an excess of electrons and the positive has too few.

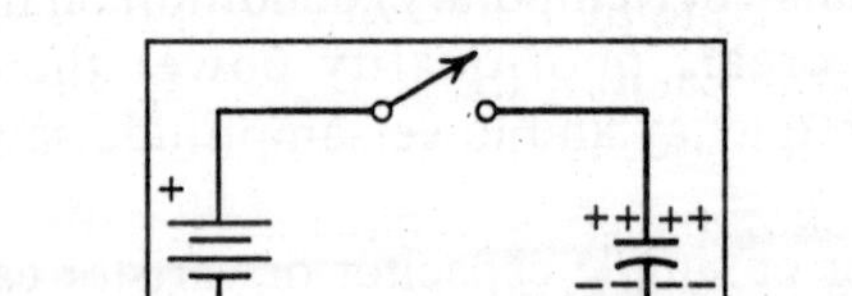

Fig. (*a*) Circuit Diagram of a Capacitor Connected Across a Battery.

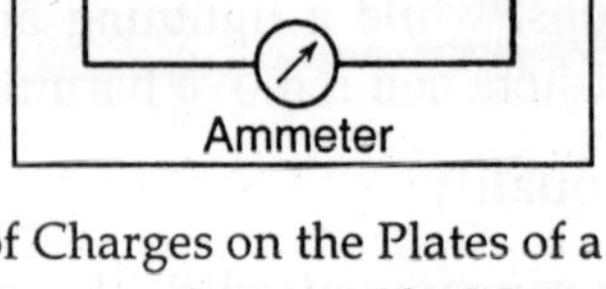

(*b*) The Distribution of Charges on the Plates of a Capacitor after it has been Connected Across a Battery. The Meter in the Circuit is a Fast, Sensitive Ammeter. It will Show a Momentary Deflection when the Switch is First Closed.

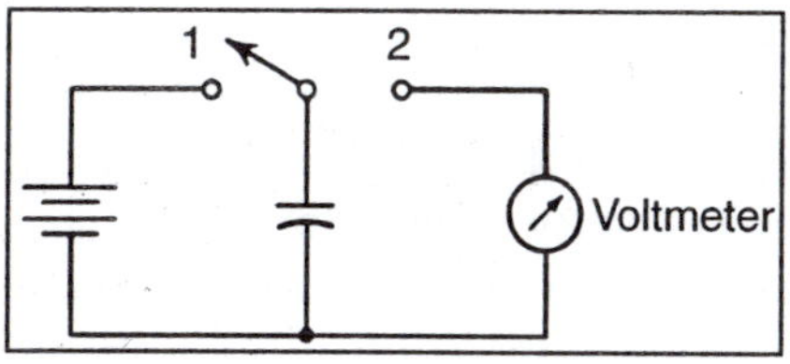

(*c*) Circuit to Indicate the Discharging of a Capacitor. When the Switch is in Position 1, the Capacitor is Charged. When it is First Thrown to Position 2, the Voltmeter will Read a Voltage (actually the Battery Voltage), and if the Switch is Held in Position 2, the Meter Reading will Gradually Approach Zero.

Since each of the plates was originally neutral, some of the electrons from plate *A* will migrate to the "+" terminal and some of the electrons from the "-" terminal will move up to the plate *B* of the capacitor. Because the glass is a good insulator, they will stop there and, when equilibrium is reached, no more current will flow. The charges will then be distributed. If a very fast and sensitive ammeter were connected to record the current as shown in the figure, it would register a short pulse of current when the connections were first made, and then no more.

Now the capacitor may be disconnected from the battery and, as long as the two leads are not connected together, the charge will remain on the plates. As soon as a path is provided for the excess electrons on the negative plate to flow to the positive plate, a current will flow through the path until the charges are again neutralized. The voltmeter provides such a path, and it will indicate a voltage (current will flow through its coil) that decreases as the charge dissipates. You can see this phenomenon easily by using the equipment on your board.

Your capacitors are of a type that works properly only when the impressed voltage is of the correct polarity. Connect one of them across the 22½ volts of your battery, making sure that the end of the capacitor marked "+" is connected to the "+" terminal of the battery and the "–" to the negative terminal. (Each of the wires sticking out of the end of the capacitor is connected to one of the plates.) Now set your meter to read at least 22½ volts d-c, disconnect the capacitor from the battery, and connect it between the meter leads. The meter will read something less than 22½ volts, and the reading will gradually drop as the current running through the meter equalizes the charges on the two plates of the capacitor.

Now connect the circuit. When the last connection is made, you may see the meter needle twitch as the current charging the capacitor passes through the meter. However, after a short time, the voltage produced by the charges on the plates of the capacitor just balances the voltage of the battery, and current ceases to flow. (If your meter continues to read something other than zero and you have connected up the circuit properly, your capacitor is faulty.) If you now disconnect the lead to the negative side of the battery and touch it to some other battery terminal, changing the voltage across the capacitor, the meter needle should jump again.

In general, the needle will read a value other than zero only when the input voltage is changing, and for a very short interval after it has changed. Furthermore, the direction in which the needle moves does not depend on the absolute voltage level but only on whether the voltage has increased or decreased. The capacitor thus behaves very much like the transformer described earlier in this chapter. It transmits changes in input voltage but passes no current when the input is steady. In other words, a capacitor also transmits the a–c component of an input voltage but blocks the d–c component.

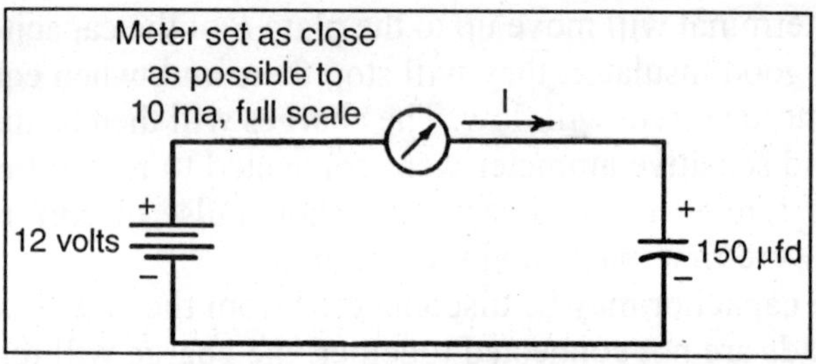

Fig. Circuit to Demonstrate the Current Flow I During the Charging of a Capacitor

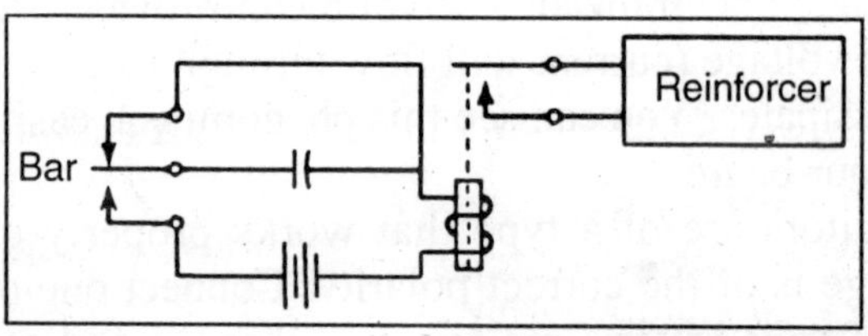

Fig. Circuit to Deliver a Single Reinforcement Each Time the Bar is Pressed. When the Switch is in its Normal Position, the Capacitor is Discharged. Upon Pressing the Bar, the Capacitor will Charge up Through the Relay Coil, Causing the Armature to Pull in Momentarily. As Soon as the Capacitor is Charged, Current will Stop Flowing Through the Coil, and the Relay will Open. When the Bar is Released, the Capacitor is Discharged, and the Circuit is Returned to its Initial Condition.

Each time the bar is depressed, the reinforcement mechanism operates, but it will not operate again until the bar is released and then depressed again. This is the general way in which a capacitor is used to detect changes in voltage. The particular values of circuit parameters (e.g., the size of the capacitor) which optimize circuit performance obviously depend upon the particular intended function of the circuit. In many cases, the experienced designer just guesses at some values, tries them out, and then plays with them until things work reasonably well.

This approach requires some understanding of the properties of charging and discharging capacitors. But it should be emphasized that expert circuit designers very often arrive at specific circuit values through crude trial-and-error procedures, because such procedures are often more efficient on the whole than trying to go through a set of elaborate computations.

TIME DELAY CIRCUITS

The term "timer," or "timing circuit," is commonly used in two different ways. It can refer either to a piece of apparatus which measures the amount of time elapsing between events (e.g., a stop watch) or to something which actually controls the time interval between events (e.g., the timer on an electric oven). This chapter will discuss only timers of the second sort.

MOTOR-DRIVEN TIMERS

A variety of motor-driven timers is available at any electronic suppliers. These timers consist basically of a set of contacts mechanically opened and closed, according to some predetermined schedule, by the action of a constant-speed motor (usually a synchronous motor). Some units, go through just one cycle and then have to be reset. The timer in this figure is designed specifically for turning on and off the lamp in an enlarger, but it can be used in many other applications as well. The timer is plugged into the wall, the gadget to be timed is plugged into the socket on the side of the timer, and the desired time interval is set on the dial.

Then when the switch in the upper right corner is turned to "on," two things happen: (1) the wall-socket voltage is connected to the timer socket (so that the enlarger lights), and (2), the timer motor starts to run, driving the pointer toward zero. When the pointer reaches zero, the voltage is disconnected from the timer socket (the enlarger goes off). The switch at the upper left reverses the condition of the contacts connected to the timer socket, so that the voltage goes off while the timer is running and turns on when the pointer reaches zero.

The timer is also an enlarger timer with an action very similar to the one just described. However, it is constructed to reset itself automatically. The pointer is set to the desired interval and the button is pushed. This connects the voltage to the timer socket and starts the motor running. When the pointer gets to zero, the voltage will turn off just as in the timer described above. But this timer will then continue through one more step, during which the pointer reverses itself and runs back to the position where it was originally set.

To get another cycle, the button is simply pressed once again. Timers such as this one, which reset automatically by mechanical means, usually complete the resetting operation and are ready to be restarted within 1 or 2 seconds, regardless of the duration of the interval actually produced by the timer. (Relay capacitor timers, such as will be discussed later in this chapter, can be made to reset instantaneously.)

As an example of the use of this timer in behavioural research, suppose that someone were interested in the question of whether or not darkness is a positive reinforcer for a nocturnal animal, say a cat. A cat is put into a lighted cage and the conditions are to be arranged such that, every time the cat makes some response, the light will turn off for 10 seconds, then go back on again. If the experimenter simply plugged the light into the socket of a resetting timer and turned the switch to the "off during the cycle" mode, his only problem would be to train the cat to make the response of pushing the button on the timer. Since this is not a very easy response for a cat, there is a better solution.

The push button is a single-pole, single-throw switch that is normally open, and which closes momentarily when the button is pushed. All the experimenter has to do is take apart the timer, connect two wires, one to each of the push button contacts, and bring the wires out of the timer casing. Now any time these two wires are connected together, the timer will go through a complete cycle. In this way, a response more suitable to a cat's behaviour repertory can be made to turn off the light.

For example, the end of the cage may be hinged and set up to close a normally open, single-pole, single-throw microswitch. Another common type of motor-driven timer runs continuously, repeating some predetermined cycle over and over. A set of cams is driven by a synchronous motor, and each of the cams presses on a microswitch.

As the motor shaft turns, the cams operate their microswitches in a sequence that can be changed by adjusting the positions of the cams with respect to the motor shaft. Similarly, the length of time that each microswitch is on or off can be changed by changing the shape of its cam.

These timers can be bought with almost any number of cams and microswitches, and with motor speeds ranging from about 2 revolutions per second up to I revolution per year. They are useful when a series of events should occur in some fixed time sequence, and the duration of each cycle is known in advance.

MISCELLANEOUS TIME DELAY RELAYS

For timing that is accurate and reliable to a few per cent, at low or moderate cost, motor-driven timers are probably the most suitable. But for applications in which the interval need not be maintained that accurately, there are other timers that are cheaper. This timer consists of a heating coil and two contacts, one of which is mounted on the end of a bimetallic bar.

When current is passed through the coil, it heats up the bimetallic bar, causing it to bend. If the unit is normally open, when the bar gets hot and bends far enough, the contacts close and then stay closed as long as current continues through the coil.

The time between the onset of coil current and the closing of the contacts is determined primarily by the physical characteristics of the bar and the coil as well as by the voltage across the coil. A normally closed unit behaves in the same way except that the contacts are closed at first, and open when the bar bends far enough. These time-delay units are relatively cheap and are available with delays ranging from about I second to 30 seconds. However, both their age and the room temperature affect the time interval, so they should not be used where the reliability of the interval is critical.

As long as current continues to flow through the heater coil of a thermal time-delay relay, the contacts will remain in the same condition (i.e., closed in a normally open relay, open in a normally closed relay). When the current is turned off, the bimetallic bar cools and the contacts return to their "normal" condition. The cooling process is slower than the initial heating. This kind of relay, therefore, cannot be used for operations which must be restarted soon after the first cycle is completed.

Timers using the viscosity of fluids to achieve a time delay are also available. They are somewhat more consistent than the heater type, and many of them can be set to give a range of intervals by adjusting the diameter of an aperture through which a fluid flows. These timers are also more expensive than the heater type. It is sometimes necessary in the design of circuits to include a short delay the duration of which need not be adjusted, e.g., to ensure that one event occurs before another. This is accomplished in some commercial time-delay relays in the following way.

A slug of copper is built into the body of an otherwise standard relay in such a position that some of the magnetic field of the relay coil passes through it. When a voltage is first impressed across the relay coil and the magnetic field begins to build up, it generates a current in the slug which, in turn, tends to oppose the magnetic field of the coil. This interaction results in a slower buildup of the net magnetic field. In this way, the time between the application of voltage to the relay coil and the closure of its contacts can be increased from the normal time (a few milliseconds) to as much as one-half second. Such a relay is called a "slow-make" relay.

By placing the copper slug in a different position, the collapse of the field that occurs when power is removed from the relay coil may be slowed down. That kind of relay closes rapidly when power is applied, but the contacts remain closed for a longer period than normal when the power is removed. It is therefore called a "slow-break" relay. Since it is not really the purpose of this book to catalogue available electric devices but rather to discuss the uses of basic principles in the design of electric apparatus, the preceding discussion is not very extensive. Furthermore, for many of the situations in which timing is required, you can build your own time-delay unit by applying some of the principles already discussed.

RELAY-CAPACITOR TIMERS

By putting together simple combinations of capacitors, resistors, batteries, and relays, it is possible to build time-delay circuits which will be sufficiently versatile and accurate to fill many of the needs that arise in behaviour research. When a capacitor is connected across a voltage source, charges build up on the two plates of the capacitor, and when the capacitor is fully charged, the voltage due to those charges is equal in magnitude and opposite in polarity to the source voltage.

When the capacitor is then disconnected from the source and the two plates are connected to each other through some conductive path, current will flow until the plates are discharged. The solid line is a plot of the current flowing out of a capacitor as a function of time when there is a path *ab*, from one plate to the other, containing a resistance *R* and an ammeter. The curve is an exponential decay curve that approaches zero asymtotically.

Consider the state of the capacitor after it has been connected across a 22½ -volt battery for a long time, then disconnected from the battery but not yet connected to the meter. There is an excess of electrons on the negative plate relative to the positive plate, and such a relation defines the fact that there is a voltage between the plates. Just before the capacitor was disconnected from the battery, the circuit was in equilibrium and no current was flowing. This means that the voltage between the plates of the capacitor was just big enough to balance the voltage of the battery itself; that is, there was enough charge on the plates of the capacitor to produce a voltage just equal in magnitude and opposite in direction to the battery voltage. When the battery is removed, and from that time until a path is provided for current to flow between the plates, the voltage across the capacitor will be 22½ volts. Therefore, at the instant when the path is first connected across the capacitor, the current flowing will be:

$$
\begin{aligned}
I &= \frac{E}{R} \\
&= \frac{22\frac{1}{2}}{100} \\
&= 0.225 \text{ ampere}
\end{aligned}
$$

This is the value of the current at $t = 0$, the first point on the solid curve. As time passes, the fact that current is flowing means that the difference in electron densities between the two plates is being reduced. Therefore the voltage between plates is being reduced, which, in turn, means that the current will be smaller. This is the condition for an exponential decay.

Now consider what would happen if the resistance in the path were to be cut in half, to 50 ohms. The initial voltage would still be the same and the initial current would be twice as great. Therefore, the charge would dissipate more quickly and the curve should be steeper.

The dashed curve is a plot of the 50-ohm condition. Similarly, the solid and dashed curves represent the *voltage* across the capacitor as a function of time for the 100- and the 50-ohm conditions respectively. The steepness of these curves clearly depends upon the rate at which electrons can flow from one plate to the other, i.e., the resistance of the circuit. But the steepness also depends upon the amount of charge that the capacitor has in the fully charged condition.

In other words, if two capacitors are initially charged to the same voltage, and both are then discharged through 100-ohm resistors, the one that was constructed to soak up the bigger charge in the first place will be longer in losing its charge because it has more to lose. For example, if the plates of one capacitor have twice the area of the plates of the other, the larger one will hold twice as great a charge when they are both fully charged to the same voltage. If they are now discharged through equal resistances, the voltage across the larger one will decay more slowly. The curves represent the voltage-decay characteristics of these two circuits.

One might expect that, if the larger capacitor had more charge on it at $t = 0$, it would show a higher voltage than the other at $t = 0$, but the voltages across the two are equal at $t = 0$. The voltage that appears across a capacitor after it has been fully charged, but before it has been discharged at all ($t = 0$), always equals exactly the voltage of the source that did the charging.

This is true by definition, because a fully charged capacitor is at equilibrium with the charging source. The voltage across a capacitor is a measure of what might be considered the charge *density* as contrasted to the total charge. Charging a capacitor to a given voltage level produces a particular charge density, but the actual total amount of charge in the capacitor will depend oil some of the physical characteristics of the capacitor itself.

The term used to indicate the relative amount of charge that a capacitor contains when it is fully charged to a given voltage is "capacitance," measured in units called farads or, much more commonly, microfarads (one-millionth of a farad). The capacitance of a capacitor depends principally upon the size and closeness of its plates. Plates that are larger and closer together give greater capacitance.

The curves look similar and are, in fact, identical. A decrease in the resistance of the discharging circuit has exactly the same effect on the time characteristics of the discharge as an increase in the capacitance of the capacitor. There is a simple way of describing the time characteristics. These curves and all voltage- or current-vs.-time curves for simple capacitative circuits have the same shape, that of an exponential rise or decay. This is the classic growth curve extremely prevalent in physical and biological systems.

The curve starts at some level and asymptotically approaches some other level. The characteristics of any simple exponential curve may be concisely described by stating the "time constant" of the curve, that is, the time it takes for the curve to get from its starting level to a level $(1/e)$ X (the asymptotic level), where e is the base of the natural logarithms, 2.7. Since 1/2.7 is reasonably close to $\frac{1}{3}$, the time constant of any such curve is the time it takes for whatever value is being plotted to get about two-thirds of the way from where it starts to where it will end up.

The shape of exponentials is such that the starting level can be taken as any level at all, and the above statement is still true. For example, the time constant is 10 seconds. It takes 10 seconds for the curve to get from 100 volts to approximately 33 volts (actually 1/2.7 × 100 = 37 volts). It also takes 10 seconds to get from 33 volts to 11 volts, from 11 to 3.7, etc. It takes 10 seconds to get from 80 volts to 27 volts, etc.

The curve is a plot of the voltage across a capacitor as it discharges through a resistor. It is a very convenient fact, and one that should be retained by anyone interested in building electric apparatus, that the time constant in seconds for this or any capacitative discharge curve equals the product of the resistance through which the capacitor is discharging (measured in ohms) and the capacitance of the capacitor (measured in farads).

Time constant (seconds) = R (ohms) × C (farads)

For example, in this circuit,

Time constant = 10,000 ohms × 1000 microfarads

= 104 × 103 × 10 – 6

= 10 seconds

Now let us reconsider the problem discussed earlier in this chapter, that of turning off a light for 10 seconds after a cat closes a switch. The 10-second time interval can be determined by the travel of a motor shaft, as with the dark-room timer, or by the time it takes a bar to heat up or fluid to pass through an orifice. But it can also be established by a purely electrical phenomenon—the discharging of a capacitor. In the previous discussion, the time characteristics of the current flowing out of a capacitor were analyzed for the circuit.

Now suppose that the 10,000-ohm resistor in this circuit is replaced by a relay whose coil resistance is 10,000 ohms, and which opens when the voltage across it drops below 33 volts. In this case, since the voltage-vs.-time curve will be the same as it was before (it is determined only by the capacitance and the resistance), the relay will pull in when the switch is first thrown into the *b* position, and stay in for about 10 seconds, i.e., until the voltage across it falls below 33 volts.

Therefore, if the cat's lamp is connected up, the lamp will go off for 10 seconds when the cat presses the bar. As soon as the cat releases the bar, shifting the switch back to the *a* position, the capacitor will be recharged and the circuit will be ready for another cycle. (This circuit only gives a full 10-second interval if the cat holds the bar down for at least 10 seconds. Circuits for momentary bar presses will be discussed later.)

To get a feeling for how this circuit operates, on your board, taking care to connect the capacitor with the correct polarity. There is a resistor in this circuit in the loop through which the capacitor is recharged. If the resistance of that path were very low, a very high current would flow at the first instant that the switch is thrown to the recharging position, and the switch contacts might be damaged. The 30-ohm resistor limits the current to a reasonable value but still allows the capacitor to charge up very quickly.

If your circuit is working properly, when you press the microswitch and hold it down, the light should go on for a short time and then go off again. The time interval marked off by such a circuit can be computed, at least roughly, from the electrical parameters of the circuit. First of all, the time constant is easy to compute. It is just the product of the capacitance of the capacitor and the resistance of the relay coil. If the resistance of the relay coil in the circuit is 10,000 ohms, the time constant is:

$$150 \times 10 - 6 \text{ (farads)} \times 10{,}000 \text{ ohms} = 150 \times 10 - 2 = 1.5 \text{ seconds}$$

This means that the voltage will drop about two-thirds of the way from its maximum value (the battery voltage) to 0 in 1.5 seconds. Since the battery voltage is 22½ volts, the voltage across the relay will be $\frac{1}{3} \times 22\frac{1}{2}$ = about 7 volts, 1.5 seconds after the switch is thrown. Another 1½ seconds after that, the voltage will be about 2 volts ($\frac{1}{3} \times 7$ = about 2). If the relay is one that opens when the voltage falls to 5 volts, the light will stay on for something between 1½ and 2 seconds.

The interval marked off by this circuit may be changed by causing the voltage to drop to 5 volts more or less rapidly. One procedure is simply to change the charging voltage. This does not change the time constant of the circuit, but, since it changes the initial voltage, the time to reach 5 volts is changed accordingly. The other means of adjusting the timing of this circuit involves changing the time constant itself. Since the time constant equals the product of the capacitance and the resistance, either of these may be increased or decreased to get a corresponding increase or decrease in the time to reach 5 volts. For example, if the capacitance is doubled, say, by connecting a second equal capacitor in parallel with the one that is already there, the time will be approximately doubled. If the net resistance through which the capacitor discharges is cut in half by connecting a resistance equal to the resistance of the relay coil in parallel with the relay coil, the time interval will be cut in half. This last means of adjusting the timing is usually the most convenient, and may be done, to give a continuous adjustment from zero up to some maximum time interval.

For a relay with a given sensitivity, for instance, one that opens at 5 volts, the higher the resistance of the relay the greater the time constant and the time interval. However, it does *not* follow that adding a resistance in series with the relay coil, will increase the time the relay is closed.

An addition of resistance equal to the relay coil resistance, for example, will indeed double the time constant of the circuit, and the voltage across the *capacitor* will take twice as long to reach 5 volts, but it is the voltage across the *relay* that matters and, since the relay and the series resistor have equal resistances, the voltage across the relay will be 5 volts when the voltage across the capacitor is *10 volts*, not 5. The capacitor voltage will actually reach 10 volts with the series resistor in less time than it would reach 5 volts without the resistor. Thus, adding resistance in series with the relay coil is just another way of *reducing* the time interval.

Now examine the circuit. Here, the capacitor is connected in series with the relay coil and the battery. When the switch is in the *a* position, the capacitor is completely discharged. At the instant that the switch is thrown into the *b* position, a current will flow through the battery and relay to charge up the plates of the capacitor.

The charging of a capacitor follows exactly the same laws as does discharging, so that the curve is again an exponential whose time constant equals the capacitance of the capacitor multiplied by the resistance of the path through which the capacitor is charging. Therefore, the relay will close when the switch is thrown to *b*, and stay closed until the charging current drops below whatever level is necessary to hold the relay closed. The action of this circuit may be conveniently diagrammed, and the duration of the relay closure may be determined from the time constant of the circuit and the relay characteristics exactly.

Most relays that are reasonably priced and suited for the sorts of timing purposes discussed in this chapter have two special characteristics that are important. First, the actual time required for closing is usually on the order of several milliseconds, so that it is not a good idea to try to build a relay timer for intervals much shorter than 25 to 50 milliseconds unless a special-purpose relay is used. Second, the highest coil resistance readily available in relays is about 10,000 ohms. Since it is also practical to buy capacitors with values up to 5000 microfarads or so, simple relay-capacitor timers may conveniently have time constants of up to

$$104 \times 5 \times 103 \times 10 - 6 = 50 \text{ seconds}$$

For a given time constant and a given relay, the higher the applied voltage, the longer the time delay. (It takes the voltage longer to drop to the relay's opening level if it starts higher.) But the time delay cannot be extended too far in this way because, at the time the switch is first closed, the full voltage is impressed across the relay, and since relays are not generally designed to withstand voltages very much greater than the voltage required for normal closing, the starting voltage cannot be too high.

As a rough rule of thumb, it is probably safe to use a voltage source about 6 times as high as the voltage at which the relay just opens or closes. Since the voltage drops two-thirds of the way in one time constant, this means that the longest delay that should actually be produced by this sort of circuit is about double the time constant of the circuit. In general, it is relatively easy and practicable to build relay-capacitor timers to operate over the range of about 50 milliseconds to 100 seconds.

The two timing circuits already discussed operate when, for example, a light is to stay on for the first 10 seconds of a steady bar press. This circuit is sometimes called a step to-pulse converter, because the input is a step and the output a pulse of fixed duration. This kind of converter is a useful building block in many more complex circuits, as will be illustrated later.

Now consider a slightly different problem that of causing the light to go on when a bar is pressed and go off 10 seconds after the bar is released. This requires that the relay controlling the light stay closed after the switch is opened. Therefore some form of holding circuit is suggested, but the holding circuit must open itself after 10 seconds. The action of this circuit is as follows. Before the bar is pressed, no current can flow anywhere because the open bar switch blocks the only path through the power source. As soon as the bar is pressed, current will flow through the relay coil, and the relay will close.

The closing of the relay connects the capacitor and the relay coil in parallel across the battery, and the capacitor will charge up through resistance *R*. (The resistance *R* here, again, serves to limit the otherwise very high current flow at the instant when the relay is first closed. Its value should be low, e.g., 30 ohms, so that it will permit the capacitor to charge up rapidly.) The relay will clearly stay closed as long as the bar is held down. At the instant the bar is released, the battery will be disconnected from the relay coil, but the capacitor will still be connected across the coil.

Since the capacitor was connected across the 22½ volt battery until the last instant before the bar was released, the capacitor itself will deliver a potential of 22½ volts. Therefore, at the first instant after the bar has been released, there will still be 22½ volts across the relay coil and the armature will remain pulled down. The capacitor will then discharge through the relay coil, and the relay will finally open when the discharging current falls below the value necessary to hold it closed. This time interval may be calculated as were the others mentioned above. Build this circuit on your board. It is an important and useful one.

A different timing requirement is one in which a delay is to be introduced between the time a response is made and the reinforcement is delivered. To take the simplest case of this kind, suppose that a human subject is to be presented with a flash of light I second after he presses a key, and he is instructed to hold down the key until after the flash has occurred. The problem is diagrammed, the flash being delivered by a camera shutter which is operated by a solenoid.

The circuit almost solves the problem. When the subject closes his key, the relay closes (through contact 1 and a capacitor). The time constant of the loop through the relay coil is chosen so that the current charging the capacitor will drop low enough to open the relay after 1 second, as diagrammed in the upper right. The solenoid is connected in series with contact 3 (on the relay) and contact 2 (on the key), so that it will operate only when both contact pairs are closed. When the key is first pressed, both of those contacts are closed, but contact 3 immediately opens (when the relay closes) and then stays open for I second. At the end of that time, contact 3 closes again and the solenoid will be activated.

The current through the solenoid can, therefore, be diagrammed. The duration of the first short pulse of current depends upon how long it takes the relay to close (and contact 3 to open) after the key has been pressed. If the relay is fast enough and the solenoid slow enough, that first pulse will not operate the shutter and only one flash will be delivered, but there is a chance that the pulse will actually trigger the shutter and two flashes will be delivered. Although it may be possible to find a fast relay and a slow solenoid to put into the circuit, it is better to change the circuit itself so that the pulse cannot occur. The pulse in the circuit occurs because contact 3 will not open until after contact 2 has closed. The circuit causes contact 3 to open before contact 2 closes.

When the subject closes his key, relay *A* closes, and stays closed for I second. The solenoid is again connected in series through contacts 2 and 3, so that both must be closed for the shutter to be triggered. When the relay closes, it opens contact 3, and that contact then stays open for 1 second, as in the preceding circuit. But, contact 2 will not close until after contact 3 has already closed. Contact 2 is initially open. Then, when relay *A* operates, it first opens contact 3 and *then* closes contact 4, operating relay *B*. When relay *B* operates, it, in turn, closes contact 2.

Now when relay *A* opens, after I second, contact 3 will close, and since contact 2 is also closed, current will pass through the solenoid. However, this circuit is not quite complete because, when relay *A* opens, it will also allow relay *B* to open, so the time during which contacts 2 and 3 are both closed will be very short. This problem is remedied in the circuit. Here relay *B* is made to hold itself closed even after relay *A* has opened.

The holding circuit is broken when the subject releases his key, and the circuit is restored to its initial state. It operates in the following way. When the key is first closed, current will flow through *R* to the capacitor and the relay. If the capacitor is initially completely discharged, it will act as a path of zero resistance (a short circuit) when the key is first closed. Therefore no current will flow through the relay at that first instant.

However, as the charge builds up on the plates of the capacitor, a larger and larger proportion of the current will flow through the relay until, when the capacitor is fully charged, it will act as a path of infinite resistance, and the current through the relay will simply equal the voltage of the battery divided by the sum of the resistances of *R* and the relay. Thus the current through the relay will gradually increase from zero to some final value, and the time it takes to reach a level great enough to close the relay will depend upon the time constant of the circuit.

The action of this circuit may be conceived of in a somewhat different, but equivalent, way. When the key is first closed, the charge in the capacitor is zero and the voltage generated in it and impressed across the relay is also zero. However, as time passes, the capacitor builds up a charge, and the resulting voltage across the relay gradually approaches some final level. When the voltage is great enough, the relay will close.

Note, in the diagram for the timing in this circuit, that the relay does not *open* until after the key has been opened. This is because, after the key opens, the capacitor will discharge through the relay, holding it closed until the discharge current falls below the relay's threshold. The calculation of the time constant for this circuit is not as simple as for the circuits discussed above, and it is not really worth going through. Trial-and-error procedures should be used for selecting the components for any particular time delay in this circuit.

The circuits only operate properly if the subject holds his key closed until the stimulus has been delivered. If he just taps the key, nothing happens. However, it is often necessary that the stimulus be delivered 1 second after a key has been pressed, regardless of the duration of the press (e.g., it is hard to instruct a pigeon to hold down his key until the reinforcement is delivered).

The requirement that the circuit continue to perform and operate the shutter even after the key has been released (should the subject give a very short press) suggests that some sort of holding circuit is needed. Relay *A* is therefore connected basically as a holding relay, the holding circuit passing through a normally closed pair of contacts on a second relay *B*, whose function will be explained later. The circuit containing the coil of relay *A* is modified slightly to satisfy the requirement that, should the key be held down steadily, only one flash will be delivered; the key must be released and then pressed again to get another flash.

This requirement is met by having the holding relay *A* closed by the current that charges a capacitor C_1. In this way, pressing the key can only have an effect for a short time while the capacitor is charging up (and the time constant of this circuit should be very short so that the charging time is short). Any time after that, but before the key has been released to discharge the capacitor, the key is essentially disconnected from the circuit.

Pressing the key closes relay *A*, and the relay then holds closed (even if the key is released) until the holding circuit is broken. A second set of contacts on relay *A* activates the circuit containing the coil of relay *B*. This relay is connected up in such a way that it will close I second after its circuit is activated.

When *B* finally closes, it fires the shutter and, at the same time, opens the circuit that was holding relay *A* closed. Relay *A* will now open, even if the key is still held closed, because the fact that the capacitor is charged up has made the key ineffective. When relay *A* opens, capacitor C_2, will begin to discharge through relay *B*, and relay *B* will soon open.

As soon as the key is released, the capacitor C_1, will discharge and the circuit is back to its initial condition, ready to repeat its operation. (That statement is not quite true. Relay *B* will remain closed for a short time after *A* has opened, while capacitor C_2 discharges through it. Therefore, if the subject presses the key again before *B* has opened, *B* will remain closed, and he will not get another flash.)

VOLTAGE SPIKE

In electrical engineering, spikes are fast, short duration electrical transients in voltage (voltage spikes), current (current spike), or transferred energy (energy spikes) in an electrical circuit.

Fast, short duration electrical transients (overvoltages) in the electric potential of a circuit are typically caused by

- Lightning strikes

- Power outages
- Tripped circuit breakers
- Short circuits
- Power transitions in other large equipment on the same power line
- Malfunctions caused by the power company
- Electromagnetic pulses (EMP) with electromagnetic energy distributed typically up to the 100 kHz and 1 MHz frequency range.
- Inductive spikes

In the design of critical infrastructure and military hardware, one concern is of pulses produced by nuclear explosions, whose nuclear electromagnetic pulse (EMP) distribute large energies in frequencies from 1 kHz into the Gigahertz range through the atmosphere.

The effect of a voltage spike is to produce a corresponding increase in current (current spike). However some voltage spikes may be created by current sources. Voltage would increase as necessary so that a constant current will flow. Current from a discharging inductor is one example.

For sensitive electronics, excessive current can flow if this voltage spike exceeds a material's breakdown voltage, or if it causes avalanche breakdown. In semiconductor junctions, excessive electrical current may destroy or severely weaken that device. An avalanche diode, transient voltage suppression diode, transil, varistor, overvoltage crowbar, or a range of other overvoltage protective devices can divert (shunt) this transient current thereby minimizing voltage.

While generally referred to as a voltage spike, the phenomenon in question is actually an energy spike, in that it is measured not in volts but in joules; a transient response defined by a mathematical product of voltage, current, and time. Voltage spike may be created by a rapid buildup or decay of a magnetic field, which may induce energy into the associated circuit.

However voltage spikes can also have more mundane causes such as a fault in a transformer or higher-voltage (primary circuit) power wires falling onto lower-voltage (secondary circuit) power wires as a result of accident or storm damage.

Voltage spikes may be longitudinal (common) mode or metallic (normal or differential) mode. Some equipment damage from surges and spikes can be prevented by use of surge protection equipment. Each type of spike requires selective use of protective equipment. For example a longitudinal mode voltage spike may not even be detected by a protector installed for normal mode transients.

Index